DiDA

D203

DIPLOMA IN DIGITAL APPLICATIONS

Book 3

Molly Wischhusen
Janet Snell
Jenny Johnson

Heinemann

Inspiring generations

Heinemann Educational Publishers
Halley Court, Jordan Hill, Oxford OX2 8EJ
Part of Harcourt Education

Heinemann is a registered trademark of
Harcourt Education Limited

Text © Molly Wischhusen, Janet Snell and Jenny Johnson, 2006

First published 2006

10 09 08 07 06
10 9 8 7 6 5 4 3 2 1

British Library Cataloguing in Publication Data is available
from the British Library on request.

10-digit ISBN: 0 435 44984 2
13-digit ISBN: 978 0 435449 84 1

Designed by Lorraine Inglis
Typeset by ✐ Tek-Art, Croydon, Surrey
Original illustrations © Harcourt Education Limited, 2005
Printed by Bath Colourbooks Ltd
Cover photo: © Stock Image/Pixland/Alamy

Acknowledgements
Every effort has been made to contact copyright holders of material reproduced in
this book. Any omissions will be rectified in subsequent printings if notice is given
to the publishers.

Post-it® is a registered trademark of 3M.

Travelbug and *travelbug.co.uk* are fictitious names and have no connection with
any company with such names at the time of printing, or in the future.

Websites
Please note that the examples of websites suggested in this book were up to date at
the time of writing. It is essential for tutors to preview each site before using it to
ensure that the URL is still accurate and the content is appropriate. We suggest
tutors bookmark useful sites and consider enabling students to access them through
the school or college intranet.

Screenshots reprinted with permission from Microsoft Corporation.

Contents

Acknowledgements

A very special thank you to Samantha Moss and Jeanette Theaker students at Croydon College, who assisted in the preparation of the Graphics Unit by producing designs for the CD Player, the MP3 Player and the lollipops and patterns used as examples of cloning. We are full of admiration for their attitude to their work and wish them all the best for the future.

BAA plc
BBC
Department of Transport
Gameplay GB Ltd
Google
Haywards Heath College (now Central Sussex College)
Jade Teo
Lloyds TSB
London Borough of Merton
Norwich Union Direct
Starfish Design and Print, Heathfield, East Sussex
Steve Holmes of Signet Construction Ltd, Eastbourne
VL Systems Ltd

Photo acknowledgements

The authors and publisher would like to thank the following for permission to reproduce photographs:

Alamy Images – page 121
Canon Images – pages 120, 121
Corbis – page 8
Getty Images/Photodisc – page 101
Logoart – pages 8, 12
Sandisk – page 65

Dedications

I could not wish for better co-authors and friends to work with than Janet and Jenny. My very sincere appreciation to Janet for holding the fort for Jenny and me when we needed it. My thanks also to Elaine Tuffery and her colleagues for their understanding and support. As always the encouragement from my family and friends is invaluable and the arrival of Sophie Grace hardly delayed me at all!

Molly Wischhusen

My thanks to the team at Heinemann for their support and especially to my special friends Molly and Jenny, my husband Bob and all my family.

Janet Snell

I dedicate this book to my beloved husband Ray who died suddenly before it was completed.

My sincere thanks go to Molly and Janet and my sons Ian and Colin and my daughter-in-law Helen, without whose support I would not have been able to finish my contribution. Thanks also to the Heinemann team for their encouragement during what has been an extremely difficult time for me.

Jenny Johnson

Introduction

Welcome to the programme of study in digital applications. You may be studying for the Certificate or the Diploma in Digital Applications. These qualifications are designed to create confident users of digital applications, who are able to apply their skills purposefully and effectively.

Which units do you need to take?

- For the **Certificate** in Digital Applications (CiDA) you will achieve Unit 1 plus one other unit of your choosing.
- For the **Diploma** in Digital Applications (DiDA) you will achieve all four units.

Each unit is the equivalent of one GCSE. The assessment for each of the units is a summative project that is set by Edexcel, the awarding body.

This book provides you with the necessary knowledge and skills to complete Unit 3.

What do you have to do to succeed?

To enable you to complete the projects, you will need to use ICT efficiently, legally and safely. You will learn all about this in Part 2, Section 1 on standard ways of working.

You will also need to develop ICT skills in a wide variety of software applications, which you then apply to complete the projects. ICT skills are developed in Part 2, Section 2: see the Skills Checklist on pages 4–7.

TiP

In preparing for your project it is essential for you to build up expertise in each of the software packages that you will be using. The project will consist of a number of tasks, each of which will require careful planning and thoughtful implementation. Therefore it is vital that you work steadily, allowing plenty of time for each task. You will find some tasks cannot be completed unless you have finished others.

In addition, you must

- learn how to plan, review and evaluate your projects
- learn what is meant by an e-portfolio and create one
- understand the topics specific to this unit.

The aim of Unit 3 is to teach you how to produce images that communicate effectively on screen and in print, and how to combine them with other components to produce graphic products.

Each final product must be fit for its intended purpose. Each of the separate activities will be given a range of possible marks. The quality and complexity of your work will determine the mark you receive for each activity.

The completed project will **not** be printed out in the traditional way. Instead you will create an **e-portfolio** that displays your work in electronic form. An e-portfolio is rather like a website where links lead the user to the different activities you have been asked to complete. The e-portfolio will be assessed by an internal assessor and an external moderator.

Features of the book

Many of the illustrations and activities in the book are centred around a fictitious company called Travelbug. Throughout the book you will find hints, tips and tasks to help you develop the skills, knowledge and understanding you need to successfully achieve your qualification.

Skills evidence – a general signpost to advise you when to save results in activities undertaken in the skills chapters, for possible inclusion in your e-portfolio

Go out and try! – practice tasks to help you develop the skills that have been described in that section of the book

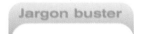

Jargon buster – explanations of some of the technical language which may be unfamiliar

Skills check – cross-references to related information in other parts of the book

Think it over – activities to get you thinking about how you will go about or prepare for the practical tasks

Tip – guidance, notes and hints to help you

Skills builder task – practice closely linked to the skills you will need to complete your summative projects (not as demanding as the full project required by Edexcel, but it will be good preparation for learning how to develop an e-portfolio)

Assessment hint – hints to guide you on what would make the difference to achieving a merit, credit or distinction compared to a pass grade

Much of the knowledge and many of the skills required for Unit 1 are also required for Unit 3. These are provided again in this book to ensure you have all the necessary information for Unit 3 to refer to in one book. A blue tab down the edge of the page denotes knowledge and skills that you should have covered in Unit 1. Knowledge and skills that were not covered in Unit 1 have no blue tab down the edge of the page.

We wish you good luck and hope you enjoy your course.

Molly Wischhusen
Janet Snell
Jenny Johnson

PART 1

Unit 3 Graphics

Introduction

Close your eyes for a short time. When you open them what is the first thing that you see? It will almost certainly be an image of something – the room you are sitting in, a view from your window, a picture in a book. We register images continuously and we rely on images to help us understand many of the things that we come into contact with every day of our lives. Can you even begin to imagine how dull life would be without the stimulation provided by images?

It is only in recent years that ordinary, everyday computer users have had the opportunity to discover the versatility of graphics software. It is no longer confined to the world of the designer or the computer expert. Today we can all be designers and use the technology of ICT to capture, create and develop images and to use them as highly effective methods of communication.

LEARNING OUTCOMES

After working through this chapter you should be in a position to

✓ use graphics software to create and edit artwork and images that communicate effectively in print and on screen.

How will I be assessed?

You will be given a project brief for this unit and you will be required to produce images and artwork for a specified audience and purpose. The project brief will specify exactly what you will be required to produce. You can expect to spend about 30 hours working on the assessment for this unit.

You will be assessed on **six** separate activities as follows:

1 Plan and manage the project
2 Select and capture digital materials from a variety of sources
3 Develop design ideas using vector-based tools
4 Develop design ideas using bitmap-based tools
5 Exhibit work in an e-portfolio
6 Review the project.

Read the project brief carefully, as it will give you hints and tips on how to complete the assessment successfully. The skills chapters provide help on the various software applications you can choose to use. In addition, you will find detailed guidance throughout this chapter, and we advise that you also refer to the following chapters, which contain specific help on completing the assessment:

- Standard ways of working
- Project planning
- Review and evaluation
- Creating an e-portfolio.

Skills file

The table below lists all the skills required to complete this unit successfully. The pages where you will find these skills explained are identified in the Page column. You should have already covered those which have a blue tab in Unit 1.

WORD-PROCESSING SOFTWARE	
	Page
enter, cut, copy, paste and move text	86
use paragraph formatting features:	
alignment	90
bullets and numbering	90
tabs	93
indents	94
format text:	
font type and font size	88
bold, underline, italic	88
colour	88

PRESENTATION SOFTWARE

ARTWORK AND IMAGING SOFTWARE

INTERNET AND INTRANETS

About graphic products

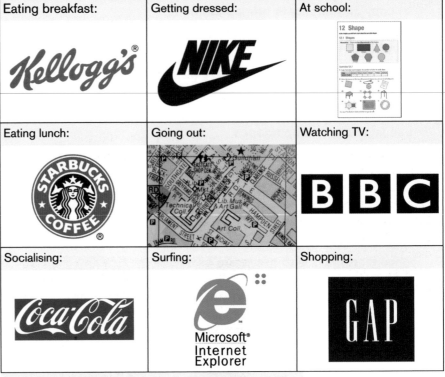

Figure 3.1 Graphical associations with common activities

It is said that 'pictures speak louder than words'. We tend to remember something more easily if there is an image associated with it. The images we are familiar with nowadays may have been produced in one of a number of ways:

- traditionally (e.g. a sketch, diagram, painting or photograph)
- digitally (e.g. captured with a digital camera or created in graphics software)
- by a combination of both (e.g. a sketch or photograph scanned and then edited in graphics software).

You see and use these images every day, for example as

- illustrations in books, magazines, newspapers and posters
- symbols and signs in public places
- symbols and icons on websites
- digital fine art on screen and paper
- brands and logos on labels and packaging
- plans, diagrams and models.

Illustrations in books, magazines, newspapers and posters

Graphics play an important part in helping to make any text more interesting and engaging, but they should be chosen with great care and thought. They should be relevant to the topic and included to serve a specific purpose, and should have been selected to suit the readers of that publication.

Consider the size of the image. An image in a poster tends to be more predominant than the text because the first thing that a passer-by will register is the image. However, the image will not detract from the written words that accompany it. Images in books and magazines are not uniform in size and they are not necessarily placed in the same position on every page. A great deal of discussion goes into the design of a publication even before the text is written, to ensure that any images used are going to enhance the overall design of the book.

Go out and try!

Your local library will have a vast range of books, magazines, newspapers and posters, all together under one roof. Visit the library to browse through a selection of books, magazines and other illustrated materials to familiarise yourself with the types of illustrations used.

1   In the children's section, look at the images in both fiction and non-fiction books and note the similarities or differences.
2   Choose a general topic that interests you, and look at books on this topic written for adults. What do you notice about the styles of images used?
3   Look at magazines written for different groups of readers – for example, car magazines directed mainly at the male market and fashion and beauty magazines for female readers. Are there any significant differences in the images used?
4   Look at the posters. There will probably be a mixture: some that have been produced professionally by national organisations and others that have been produced in-house. Look at the size of the images used and consider whether they support the message? Do the in-house posters display a particular house style?

Symbols and signs in public places

The use of signs and symbols enables a message to be simplified and presented in a familiar form. Many signs and symbols are universally recognised – which is of great benefit in an age when people travel far and wide. Look at the two examples in Figure 3.2. The designs are quite simple and uncluttered, making them easily recognisable.

Figure 3.2 Two familiar public signs that inform

The triangular design of the road sign tells us that it is a warning sign. The simple black line and the arrowhead indicate that the road bends. It doesn't take any time to understand the implications of the sign, which is vital if you have only a split second to register and react to the image.

A very simple line drawing has been used to symbolise the concept of disability. It is used the world over to identify reserved parking bays, access and exit routes in buildings, designated spaces for wheelchairs on public transport, and so on.

Symbols and icons on websites

Just as symbols in public places are standardised for ease of recognition, so are the symbols on many websites. Examples are the padlock symbol that indicates you are using a secure website, and the shopping basket icon that is used increasingly on many websites that sell products or services online (Figure 3.3). Both of these images leave the user in no doubt about their purpose because relevant symbols have been chosen.

Figure 3.3 Two website symbols

You will find a wide selection of buttons and icons available on the Internet that you are free to copy and use without charge and without infringing copyright laws. Examples are shown in Figure 3.4, which came from CoolArchive (see www.heinemann.co.uk/hotlinks (express code 9842P) for more).

Figure 3.4 Examples of free website icons

Navigation links are a vital element of good website design and there are many innovative buttons in use. There are many websites that also let you download buttons for free. Figure 3.5 shows a selection of buttons that can be downloaded free of charge from Freebits (see www.heinemann.co.uk/hotlinks (express code 9842P) for more).

Figure 3.5 Examples of free website buttons

Digital fine art on screen and paper

Digital art is a relatively new art form; it began in the mid-1950s when computer programmers started experimenting with visual images. The widespread introduction in the 1980s of graphics software, such as Microsoft Paint, which didn't require programming skills, encouraged artists and designers to put down their paint brushes and experiment with computer art – and it has gradually evolved to the high-tech artwork we see today.

You will see on your television screen a lot of digital art, such as backgrounds to music videos on the music channels, introductions to programmes, advertisements and cartoons. Creators of digital fine art find it more difficult to display exhibitions of their work than do artists who still use brushes and canvas. However, you will find some excellent examples of digital art on the Internet, and in particular on the website of the Digital Art Museum. This is an online resource for the history and practice of digital art, exhibiting the work of many artists. The website can be accessed via www.heinemann.co.uk/hotlinks (express code 9842P).

If you look carefully at items such as posters, CD and book covers, and greetings cards you will find many examples of digital art in print. As its popularity develops, more and more books are published about this art form.

Brands and logos on labels and packaging

Companies spend hundreds of thousands of pounds on their brand images and logos (Figure 3.6). The brand is the 'trademark' of the company. For example, Pepsi and Pizza Hut are household names that are easily identified by their logos and product brands. The logo provides the key to establishing a corporate identify and raising its profile.

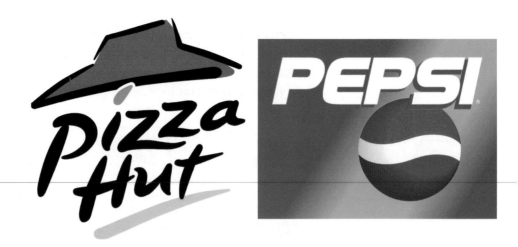

Figure 3.6 Examples of brands and logos

A logo is often produced in corporate colours, and the same colours will be reflected on company documentation, merchandise and the company's website. Companies want something that is clear, simple and easily recognised. Simple shapes or line drawings are often used to symbolise the product.

For example, a logo representing a food or drink company might simply include a 'whiff of steam' emerging from one of the letters in the company name. A company offering a wedding service might incorporate a heart as part of the letter W in Wedding.

A logo is used to promote the company and will appear on everything the company produces. If you walk through any supermarket you will recognise many products by their packaging and brand image. Think of some well-known breakfast cereals and chocolate bars for example. Companies spend millions of pounds on package design to make sure that their products stand out from the rest.

Clothing designers will market their products through their brand name, and will hang labels from the garments to promote their image as well as sewing labels inside. Some garments even have the label on the outside as a design statement! Most designers will also have their own distinct carrier bags that also carry the logo or brand name.

Airlines incorporate their logo on the tail fins of their fleet, travel tickets, luggage labels, brochures, merchandise, uniforms, and so on.

The website LogoLounge (see www.heinemann.co.uk/hotlinks (express code 9842P)) is a showcase of the work of many top designers throughout the world and is an interesting site to browse for ideas.

Some brand images are considered to be a status symbol. Burberry is a brand that is instantly recognisable by its distinctive check pattern, and handbags carrying the Louis Vuitton logo are very exclusive fashion accessories.

Go out and try!

Sapheena is going to open a hairdressing salon. She is thinking of calling it something like 'Shape'. She would like a logo designed that she can put up above the new shop. Work with a group of your friends and brainstorm ideas. Can you come up with a more appropriate shop name? Produce a *simple* sketch design for a new logo based on the name you choose.

Plans, diagrams and models

An architect prepares detailed plans that guide the builder through the construction stages of a building project. The plans show the size and shape of the building, the position of doors and windows, and construction details such as sizes of roof supports and colours of bricks and roof tiles. Site diagrams show the position of the building relative to surrounding landmarks (Figure 3.7).

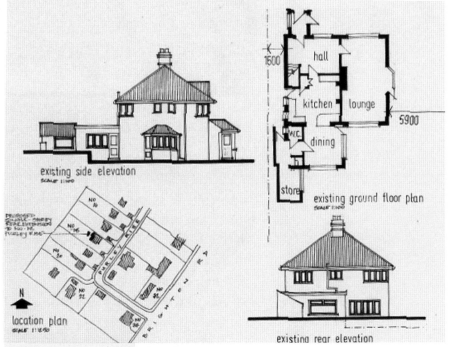

Figure 3.7 A house plan

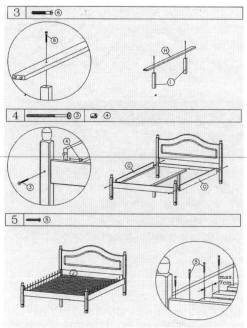

Figure 3.8 An instruction manual

Diagrams are often used to simplify detailed descriptions of complex processes. Textbooks contain diagrams illustrating topics such as the human digestive system, or flow charts illustrating the navigation routes through websites. Most flat-pack furniture requiring self-assembly will come with diagrams showing how the furniture is to be assembled (Figure 3.8).

Models are often produced to help people to visualise a finished product. This may be the prototype of a new product, or a model of a proposed town centre development scheme put on display in the town hall. However, the time and cost involved in producing models is enormous, so many designers now use computer-aided design (CAD) software to transform a two-dimensional image into a three-dimensional (3D) visual model of the product. They can then work with the digital model and try out different colours and effects before putting the product into production.

Go out and try!

Collect examples of a variety of graphics, logos, brand images, interesting font styles and other design features, from packets and magazines that you have at home. Look for interesting graphics on leaflets in high street shops. Ask at home if you can look at examples of logos on business letters. Keep a record (notes or sketches) of any interesting graphics you see on advertising hoardings.

Evaluating the effectiveness of images

To be able to design your own successful graphic images in the future, you must learn how to evaluate the effectiveness of images in relation to

- purpose
- composition
- use of colour
- impact/visual effect
- size and position
- message.

Purpose

Published images are usually created for a specific purpose:

- to express an idea
- to give directions
- to tell a story
- to convey a message
- to sell a product.

Some images will be designed to stand alone and some will be supported by text.

We look at illustrations in magazines to get ideas for fashion or interior design trends. We use maps and diagrams to find and give directions. If the purpose of the image is to tell a story, it may be necessary to create more than one image or even a whole cartoon strip. Images whose purpose is to convey a message or sell a product must make a visual impact that will hold the attention of the intended audience.

Whatever the purpose, an image is a powerful communications tool and the right choice of image is therefore very important.

Composition

Whether you are drawing a sketch, taking a photograph or creating an image in graphics software, you must give some thought to the composition of the graphic, both in terms of overall appearance and how the image will be put together.

Consider the position of the main elements in relation to each other and how the graphic will be built up. For example, if you were taking a digital photograph you would not normally position a person in front of a tree if it would appear that the tree was growing out of the person's head!

If the purpose of your image were to illustrate a recipe, you would have to consider many things when planning the overall composition of the image:

- Will the image be in black and white or colour?
- Will the image of the food be in the foreground?
- What would make an appropriate background?
- What background colour would enhance the food rather than overwhelming it?

- What size is the image intended to be?
- Might the overall image be enhanced by using accessories (such as a vase of flowers, a jug or a glass of wine)?
- Will text be layered over the image? If so, would your choice of background lend itself to the addition of text, or does the background image contain too many contrasting colours and textures?
- How will the image appear in relation to other images on the page or screen?
- Does the image have to fit in with an overall colour scheme?

The image in Figure 3.9 is a digital photograph, taken by an amateur photographer late on an autumn day to capture the effect of the clouds and sun against a church standing on a clifftop. The bright sky appears to outline the church. It is a simple photograph but the contrast of the dark church to the bright sky adds some interest. Imagine this image with a text overlay. It would work well because the intensity of the sky is fairly even and text in a dark colour would stand out. Similarly, text at the foot of the picture could be presented in a light colour to contrast with the dark shade of the fields.

Figure 3.9 A digital photograph

Use of colour

You must consider an image's purpose and where it will be published before you can decide whether or not to use colour.

Black and white can create a strong visual identity and can be more economical to reproduce on paper than colour. Sometimes an image on a white background can be far more effective than an image on a coloured background.

When you investigate logos, you will probably notice that their designs rely on very few colours, whereas other art forms use many more colours. Background colours applied to screen-based publications can increase their effectiveness and provide you with an opportunity to use contrasting colours for the text. Dark backgrounds on screen, in particular, can be very effective.

You should also consider whether paper-based documents could be printed on coloured or textured paper, or even on acetate.

Impact/visual effect

If the purpose of an image is to sell a product or convey a message, it must make a visual impact to persuade the audience to stop and take notice. To attract the public to buy something, an image must depict the product in a very positive way. Happiness, laughter, sunshine, fun and bright colours are all positive elements.

Charities often use simple images to create impact. For example, during the Comic Relief appeal we see faces of children, often with sad, tear-filled eyes. Quite often the images are in black and white to add to the overall sombre effect. The images presented are designed to stir our emotions, and require little or no text to support them.

Graphics software can be used to achieve visual effects such as blurring, rippling, texture and crayon. You will learn how to apply these as you begin to develop your own images. Using a background or theme is another successful way of adding impact to an on-screen publication.

Figure 3.10 This poster was used by the London Borough of Merton in a recent anti-graffiti campaign to achieve the maximum impact

Size and position

The size of an image must reflect its purpose. A diagram or map, for example, must be large enough for the details to be read without a magnifying glass. If an image doesn't include fine detail then the size may not be so critical, although the position on the screen or page may influence the final size. Images on posters must be visible from a distance, and images used on company notepaper, business cards, labels and other stationery should not be overwhelming. Sometimes an image used as a watermark can also be an effective solution.

The message

Sometimes a picture alone can be enough to make a statement. At the time of the Tsunami disaster at the end of 2004, one picture in particular seemed to sum up the whole tragedy. It was of a father carrying the body of his child. No words were necessary to understand the suffering that he and the thousands of other survivors were having to face.

Your skill will be to ensure that the images you use either support the message you are trying to get across or stand alone in their own right without the need for words.

Go out and try!

1 Look through images you have collected and choose one that you think is particularly effective. Look at the images your friends have collected. Compare and evaluate the images in terms of purpose, composition, colour, impact, size and message.
2 Choose two or three images that you all agree to be the most effective. As a class, produce a list of the elements that made the images successful.
3 Make a copy of the list so that you can refer to it when designing your own graphics.

Selecting and capturing digital materials

Inspiration and ideas

Most people need something to inspire them with ideas, and one source of inspiration comes from looking at work created by other

Assessment Hint

You will be asked to produce evidence that you have collected a variety of stimulating material from both primary and secondary sources, and you must present this collection as evidence in your e-portfolio. This collection will give you ideas that will be reflected in your own graphic creations. The sooner you start to build a collection of examples, the more you will have to refer to when you complete your project.

TiP

Remember to save your files in the correct format for print or screen (see Figure 3.48 on page 49).

people. The wider the range of graphic styles you experience, the more ideas you will have to apply to your own designs. It is essential that you start collecting examples of a wide range of graphic images as soon as possible so that you build a library of ideas. You are likely to draw your ideas from

- photographs or parts of photographs
- sketches, drawings or paintings of people, places or objects
- diagrams, maps and plans
- background images or textures
- text of a particular font
- unusual colours, patterns or effects you would like to recreate.

Look *critically* at graphics you see around you. If you come across a graphic or an idea that you like, add it to your collection. In fact, look at your own computer screen, which is full of graphic images designed to represent features or actions.

- Material from *primary sources* is likely to come from your own drawings, sketches or photographs and from images you have taken using a digital camera.
- *Secondary sources* will include photographs taken by other people, books, newspapers, magazines, collections of images on CD–ROM or material found on the Internet.

Some of this material will be in the form of paper-based publications, and you will have to scan these images to include them in your collection.

It is always wise to check that the images you collect and save in your catalogue appear on screen as you intend.

- Occasionally an image file is corrupted. When you open the file, part of the image may be damaged or even missing.
- If you are scanning an image, diagram or map, check to make sure you have scanned the whole image and that the map or diagram is complete. It is very easy for a document to slip or move slightly when the scanner cover is closed.
- When you are using a digital camera, don't just rely on the LCD panel on the camera to check the detail. It will be more reliable to view the larger image on the computer screen.

We have already suggested one or two websites that will provide some excellent source materials, but you must remember the laws of copyright. *Do not* use any materials or images that are protected by copyright without first obtaining the permission of the person who

Skills check ⟫

To add items to your collection you will need to be able to

- scan images
- import files
- copy and save images
- upload images from a digital camera.

These skills are covered in the artwork and imaging chapter on page 119.

✔ **TiP**

*You might choose to use an **image database**, such as Adobe Album or Picasa, to organise your image files. Image databases give you a thumbnail view of your images, and make searching easier. You may even be able to store information as notes or comments against an image.*

Skills check ⟫

Look at the artwork and imaging skills chapter (page 119) for information on vector and bitmap images. On the following pages you will discover more detailed information about these graphic types.

owns the copyright (usually, but not always, the person or organisation who produced the work).

Examples of work in the Digital Art Museum are protected by copyright. However, the copyright notice states that you may print, store or download information from the site *for your personal non-commercial use*. This means that you can store copies of work as reference materials. It is important to check all copyright notices carefully, because this right may not apply to other sites.

Organising and storing your graphics

You will need to devise a system for storing graphic files in a logical order so that you can find them easily at any time. It is sensible to do this as soon as you start to save the first few images. If you leave it, your files will get into such a muddle that you won't be able to find anything without wasting a lot of time. Devise a structure with one folder for each of the different types of graphics that you intend to collect.

You must be able to acknowledge the sources of all the materials you use. Keep a word-processed catalogue which records the following:

- graphic title or description
- artist/creator
- graphic type (font style, clip art, logo, diagram, digital photo, texture, web button, etc.)
- source
- date acquired
- file saved as ...
- file location (folder/sub-folder)
- file type (JPEG, bitmap, vector, etc.)
- copyright details.

If you record this information in a table, you will be able to sort the data to find what you are looking for. This will be especially useful when you have built up a large collection in your graphics library.

Developing images

To prepare for your assessment, you will spend a considerable time investigating the various images you will read about in this chapter. The software packages you will use have a variety of tools and techniques. You will be able to recreate (i.e. copy) some of the designs of other artists by experimenting with these tools and techniques in both vector and bitmap software, and with the aid of a digital camera. You will be able to look at alternative ways of achieving the same effect, and you will gradually develop a style of your own. At the same time, you will explore a variety of ways in which you can deliver your message.

As you create and save your images, it is essential to store each one in an appropriate *digital format*, taking into account whether the image is intended for print or screen. The summative project brief will identify the file types you must choose for the project, but as you prepare your practice material you must also give this some consideration. Here are some questions you should ask yourself:

- What size and quality is required?
- Is colour needed?
- What resolution should be used?
- What file format should it be saved in?

We have already looked at the basics that combine to produce an effective image:

- purpose
- composition
- use of colour
- impact/visual effect
- size and position
- message.

You must remember to keep these attributes in mind as you develop your own ideas, in order to ensure that each image meets its specific objective.

Skills Builder Tasks 1 and 2

At this stage you should be able to tackle Tasks 1 and 2 of the Skills Builder mini project on page 56. Don't forget to study the scenario carefully so that you are clear about the project objective.

Your assessment project

When your teacher or tutor feels that you have developed your skills sufficiently, you will carry out the summative project for Unit 3.

The assessment project will be presented as a short scenario. After studying the scenario, you will be able to identify the specific project objectives and produce a list of the tasks to be completed.

The task list will include a series of effective graphic images that you must produce in order to meet the objectives of the project brief. Some of these images will be selected from secondary sources, such as books or the Internet, while others will be from primary sources, created by you using vector and bitmap software or captured with a digital camera or scanner.

Your project will be looked at by an assessor and a moderator. They will be looking to see how effectively each of your images meets its objectives and suits its purpose (what each image is meant to do), which might be to

- express an idea
- give directions
- tell a story
- convey a message
- sell a product.

In addition they will be looking to see whether your images effectively address the following questions:

- Who is the target audience?
- What medium is your image for?
- Where will it be displayed?

Developing design ideas using vector-based tools

A vector image is made up of a collection of independent objects – lines, circles and squares – each of which can be individually selected, moved, resized, filled with colour and edited. As a result, they are easier to edit than bitmap graphics. You can draw a circle 2 cm in diameter and then resize it to 4 cm in diameter (Figure 3.11), or even 100 cm – the quality of the image will be retained.

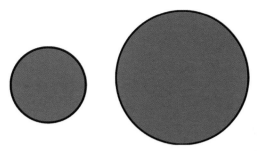

Figure 3.11 Vector-based images can be enlarged without losing quality

Vector images are ideal for maps and technical drawings, where precise line detail is required, or images that do not require the full photographic colour range (such as clip art, lettering and logos).

There are many specialist graphic software programs available: some examples are CorelDraw, Corel Photo-Paint, Printshop Pro, Adobe Photoshop and Photoshop Elements (which is a cut-down, cheaper version of Adobe Photoshop). If you do not have access to one of them, you can still create very imaginative artwork using the drawing tools in Microsoft Word.

Drawing lines and curves

The drawing tools in Word provide an extensive selection of lines and shapes. Using the **Line Style** tool ☰ or **Arrow Style** tool ⇄, you can draw lines in a variety of thicknesses and formats, with or without arrows (Figure 3.12). In addition to straight lines, there are freehand

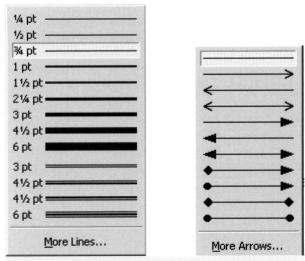

Figure 3.12 A choice of line thicknesses and arrow styles

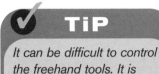

lines available through the *AutoShapes* menu: the **Curve** [ऽ], **Freeform** [⌂] and **Scribble** [✎] tools (Figure 3.13).

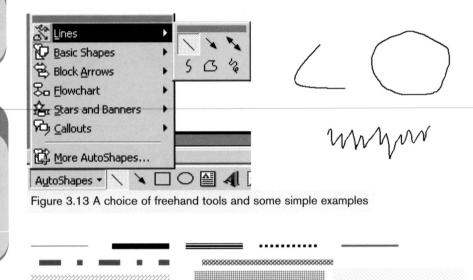

Figure 3.13 A choice of freehand tools and some simple examples

Figure 3.14 Styles of lines and borders

To draw the line or shape you require, select its icon from the menu, click on the drawing area and drag with the mouse to the size you need. If you wish to be more precise, you can give exact measurements through a dialogue box (select **Format, Format AutoShape** and then click the **Size** tab), as shown in Figure 3.15.

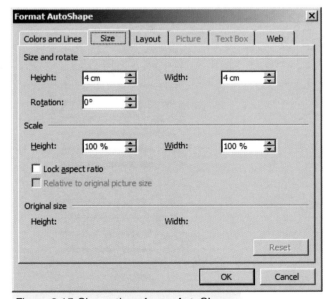

Figure 3.15 Size options for an AutoShape

Go out and try!

1 Experiment with using the various styles and thicknesses shown in the illustrations.
2 Using the line tools, draw the lines shown in Figure 3.16 – make them 2 cm wide. Experiment with a variety of line thicknesses and styles.

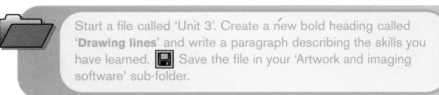

Figure 3.16 Try to create these lines to the size stated

3 Experiment with using the **Curve, Freeform** and **Scribble** tools.

Start a file called 'Unit 3'. Create a new bold heading called '**Drawing lines**' and write a paragraph describing the skills you have learned. Save the file in your 'Artwork and imaging software' sub-folder.

Drawing basic shapes

As well as the rectangle and oval available on the *Drawing* toolbar, AutoShapes provide a wide range of basic shapes: block arrows, flow chart symbols, stars, banners and callouts (Figure 3.17).

TiP

If you want to draw a square or a circle, select the **Rectangle** □ or **Oval** ○ tool, but, in order to keep the size equal in each direction, hold down the keyboard **Shift** key as you draw.

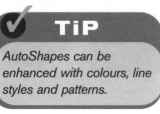

TiP

AutoShapes can be enhanced with colours, line styles and patterns.

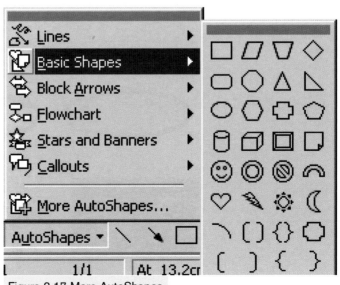

Figure 3.17 More AutoShapes

Using colour, line styles and patterns

The **Fill** tool provides numerous options for filling objects. Figure 3.18 shows the dialogue box in which you can choose a fill type (gradient, texture, pattern or picture) and the colour or colours to use.

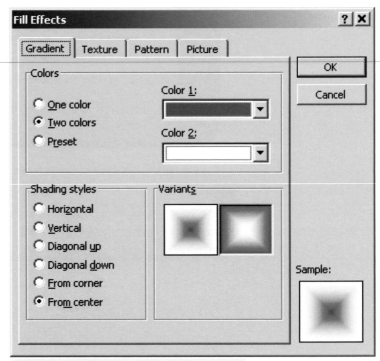

Figure 3.18 The dialogue box for Fill Effects

In Figure 3.19 the three ovals were drawn with a black outline, 4.5 points thick. One is filled with a solid yellow, the second with a pattern and the third with texture. Figure 3.19 shows also how pictures can be used to fill shapes.

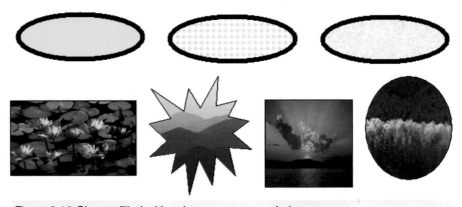

Figure 3.19 Shapes filled with colours, patterns and pictures

Go out and try!

Recreate the various shapes used to illustrate the fill effects of colour, pattern, texture and picture in Figure 3.19. You can use any pictures.

Open your file 'Unit 3'. Create a new bold heading '**Fill effects**' and write a paragraph describing the skills you have learned. Save the file.

Applying flat and three-dimensional effects

Objects can be further enhanced by using flat effects such as shadow ▣ or three-dimensional ▣ effects. Figure 3.20 shows some examples – an arrow enhanced using shadow, and a sun and heart filled with colour and shadowed.

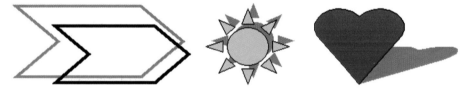

Figure 3.20 An arrow enhanced using shadow, and a sun and heart filled with colour and shadowed

Figure 3.21 shows three shapes to which different 3D options and fills have been applied.

Figure 3.21 Three shapes with 3D effects applied

TiP

*If you want the same shape again, highlight the shape and press **Ctrl+D**. You will then have a duplicate that you can move or change. Try this when creating the 3D images of the rectangle, square and oval.*

TiP

*You can move a shape or image by left-clicking on it and dragging it to a new position, or by using the arrow keys to move it up, down, left or right. Sometimes you want to move the shape or image very slightly. If you click on the image and hold the **Ctrl** key while you use the arrow keys, the image will move by tiny amounts. This can be really useful if you need to position an image precisely.*

Go out and try!

1 Copy the various shapes used to illustrate flat and 3D effects.
2 Using the AutoShape tools, draw the shapes in Figure 3.22 to the exact sizes shown, and experiment with fill colours and the line thicknesses or styles.
3 Draw at least three other shapes of your own choosing. Experiment with using different fill, flat and 3D effects.

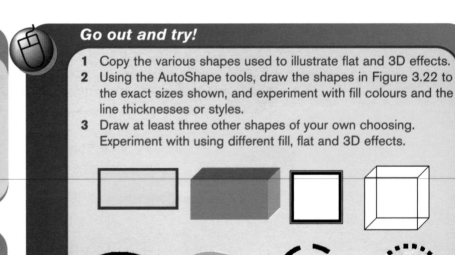

Figure 3.22 Try to create these shapes to the sizes stated: a 2D and 3D rectangle 1.5 cm by 3 cm; a 2D and 3D square 2 cm by 2 cm; an oval 2 cm in height by 3.5 cm in width; a circle 3 cm in diameter. Copy the oval and change to a 3D effect. Copy the circle and change the outline and fill effect. Draw the smiley face at any size.

Open your file 'Unit 3'. Create a new bold heading '**Flat and 3D effects**' and write a paragraph describing the skills you have learned. Save the file.

Inserting, formatting and editing text in vector images

Using a drawing package such as CorelDRAW, text can be added to a vector image as *artistic text* or *paragraph text*.

- Use artistic text to include short lines of text to your document, especially if you want to use special effects.
- Use paragraph text where you need to include large amounts of text, for example when producing newsletters, brochures or flyers.

Artistic text

To add artistic text, select the **Text** tool  and click anywhere in the document where you want the text to appear. The text becomes an object in the file and can be moved around on the page. Figure 3.23 shows an example.

Drag a handle to resize the text

Figure 3.23 An example of artistic text, and sizing handles

The text can be formatted just as you would in a normal text document: you can change the font and apply effects such as bold, italics and underline. The font size can be changed by dragging the handles of the text object, as shown in Figure 3.23.

Paragraph text

To add paragraph text, select the **Text** tool, click anywhere in the drawing window and draw a frame. Write the text in the frame (Figure 3.24).

Draw a text frame and write in the frame.
The frame can be a fixed size or can
expand or shrink to fit the text

Figure 3.24 Paragraph text added in a frame

✔ **TiP**

In Word, you can add text using WordArt or a text box.

If the frame is too small then you can increase its size by dragging on the handles. In a drawing package such as CorelDRAW, you can also allow the size of the frame to expand and shrink to fit the text, by selecting this through **Tools, Options** and ticking the box (Figure 3.25).

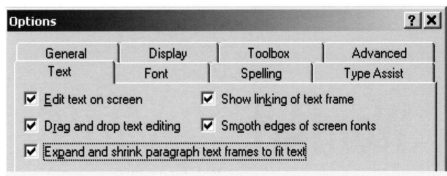

Figure 3.25 Select the option to expand and shrink the paragraph text frame to fit the text

Applying shadows and special effects

The formatting of text can be enhanced in numerous ways: shadow, outline, rotate and perspective. On selecting **Format, Font** and the **Text Effects** tab, Word even enables you to create movement or shimmer in text (Figure 3.26).

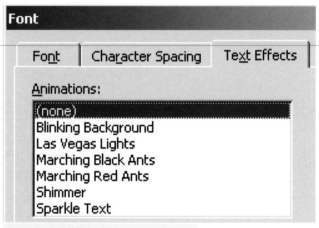

Figure 3.26 Text effects (animations)

Figure 3.27 shows some of the many possibilities for creating attractive text-based graphics. For example, the 'Perspective' image was obtained by selecting **Add Perspective** and then dragging on one of the corner nodes to stretch the text. The 'Rotate' image was obtained by double-clicking on the text to get the rotation handles, which were then dragged clockwise.

Figure 3.27 Examples of text effects

Fitting text to a path

You can position text along the path of a graphical object. This is illustrated in Figure 3.28 by the images of two globes promoting Travelbug, which could become an effective part of a flyer or other promotional material.

Figure 3.28 Wrapped text to fit a shape

The text can be placed exactly on the line or, as in this case, slightly away from the line. The line for the path was created by drawing a circle exactly matching the size of the globe, but hidden by selecting 'No colour' once the text was in place.

Go out and try!

1 Experiment with creating text using the following effects: shadow, outline, perspective, rotate, shimmer, sparkle and any other effects of your own choice. In each case, write the word and then add the effect.
2 Draw a circle and, using **Fit to Path**, add the words 'Around the World'. Add a suitable image of the world to the circle.

Open your file 'Unit 3'. Create a new bold heading **'Text effects'** and write a paragraph describing the skills you have learned. Save the file.

Combining basic shapes and freehand drawing

By combining various lines and shapes it is possible to create some interesting drawings in Word, even though the drawing tools are simple compared to the facilities provided by specialist graphics software.

Computer-aided design (CAD) software is used by companies to plan layouts for new kitchens, bathrooms or bedrooms, and at a touch of a button they can give you a 3D image of the room. If you did not have this software available, and you wanted to plan a new bedroom for yourself, you could prepare a basic plan in Word.

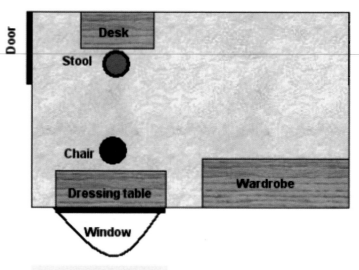

Figure 3.29 A room plan

Editing and arranging vector graphics

The plan in Figure 3.29 was created using lines and shapes which were grouped and ungrouped several times. At one stage the bed (see Figure 3.30) was very slightly outside the edge of the room, and the desk was so close to the doorway that it would not have been possible to open the door! The ungroup facility made it very easy to correct these errors, so that the bed could be moved and the desk reduced in size, after which the various sections of the plan were regrouped.

Order

When using the drawing tools, sometimes you find that one object hides another and the order in which they are *layered* may need to be changed. For example, when grouping a large number of objects, it is very easy to miss one or two. After the other objects are grouped, the ones omitted will be hidden. This is exactly what happened when the various objects in the bedroom plan were grouped – the rectangle for the bed was missed from the group and disappeared from view. This was easily corrected by selecting the plan layout showing at the front and choosing **Send to Back** from the **Draw, Order** menu; then the bed became visible again (Figure 3.30).

Think it over ...

Study the plan shown in Figure 3.30 and list all the shapes, lines and fill effects that have been used.
Check your list against the answers given on page 36.

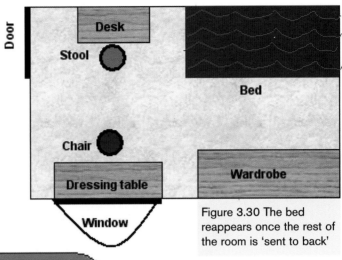

Figure 3.30 The bed reappears once the rest of the room is 'sent to back'

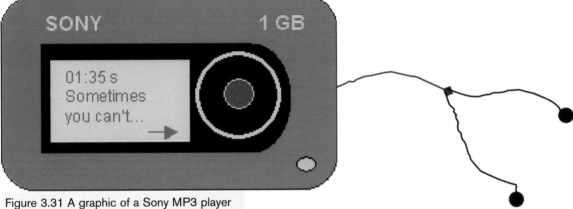

Figure 3.31 A graphic of a Sony MP3 player

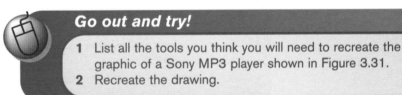

Go out and try!

1   List all the tools you think you will need to recreate the graphic of a Sony MP3 player shown in Figure 3.31.
2   Recreate the drawing.

Open your file 'Unit 3'. Create a new bold heading '**Vector drawing of an MP3 player**' and write a paragraph describing the skills you have learned. 💾 Save the file.

Copying and cloning

Cloning is an alternative way of copying an image. The difference between copying and cloning is that any changes made to the original will be updated in the cloned image.

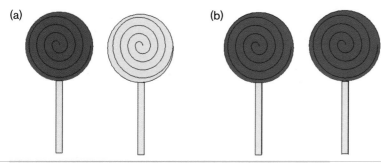

(a) (b)

Figure 3.32 (a) A copied lollipop. (b) The same image cloned

Look at Figure 3.32(a), a design of lollipops. The original colour was yellow. When the colour was changed to red, the copied image remained as yellow but the cloned image automatically changed to red, as in Figure 3.32(b). This technique can be very useful if you are designing fabric or wallpaper with a repeating pattern.

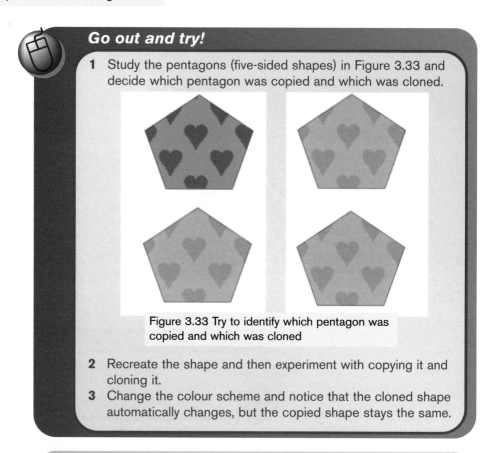

Go out and try!

1 Study the pentagons (five-sided shapes) in Figure 3.33 and decide which pentagon was copied and which was cloned.

Figure 3.33 Try to identify which pentagon was copied and which was cloned

2 Recreate the shape and then experiment with copying it and cloning it.
3 Change the colour scheme and notice that the cloned shape automatically changes, but the copied shape stays the same.

Open your file 'Unit 3'. Create a new bold heading '**Cloning images**' and write a paragraph describing the difference between copying and cloning an image. Save the file.

Combining and breaking apart objects

Objects can be merged with each other or with the background. The difference between combining objects that make up an image, rather than grouping them, is that once you have combined the objects they become one object, rather than separate objects that have been simply connected together. Once you are satisfied that the image is exactly as you want it, combining the objects has the considerable advantage of reducing the file size and so increasing the speed at which the objects can be downloaded.

Jargon buster

Native formats are those specific to the application in which they were produced. Examples are CDR in CorelDraw and MIX in Microsoft PictureIt and PhotoDraw.

Common file formats are those that are supported by most operating systems and graphics applications. Adobe Illustrator (AI) is one of the most common vector-based formats that can be used by other vector programs. Windows Metafile format (WMF) is also supported by most other operating systems.

TiP

Before converting a vector file's format, it is wise to make a copy that retains the original format. Once it has been converted it loses the benefits of vector graphics – primarily that the image can be enlarged without reducing the quality.

Skills check ▷▷

Look at page 129 in the skills section on artwork and imaging, which explains more about file formats for different purposes.

Skills Builder Task 3

At this stage you should be able to tackle Task 3 of the Skills Builder mini project on page 56. Don't forget to study the scenario carefully so that you are clear about the project objective.

Save vector images in appropriate formats

Formats for print and web

To use a vector image on a website, it must be converted to bitmap format.

To convert the file's format, select **File, Export** and choose **Windows bitmap** format. You can retain the original size. However, if you then decide the image is too small, you will need to open the *original file* and repeat the conversion process, using a higher ratio of pixels or increased resolution.

Go out and try!

1 List the vector-based software applications that are available to you at school/college or at home. Identify their native file formats, and which allow you to save files in common file formats that can be opened by different software applications.

2 Copy one of the vector files you have created and export it to bitmap format using a resolution of 75 dpi (dots per inch). 💾 Save it as 'Export 1.bmp'.

3 Open the file in Paint and, using the zoom facility (**View, Zoom**), increase the size to 200 per cent or larger. Notice how the image becomes less sharp.

4 Copy the same file again. This time, when you export the file to bitmap format, increase the resolution to 150 dpi. 💾 Save it as 'Export 2.bmp'. Notice that the second image takes up much more disk space than the first.

Open your file 'Unit 3'. Create a new bold heading '**Exporting vector files to bitmap format**' and write a paragraph describing the skills you have learned. 💾 Save the file.

? Answer to **Think it over ...** on page 33

Figure 3.34 shows the shapes and fill effects that were used to create the plan of the room.

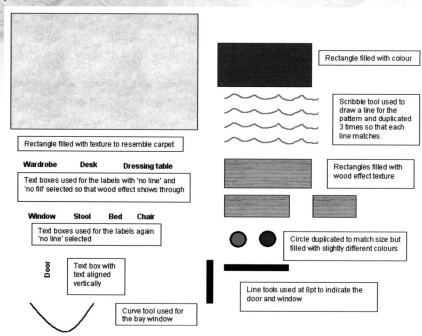

Figure 3.34 The answer to the question about Figure 3.30 on page 33

Developing design ideas using bitmap-based tools

Bitmap graphics are best suited to photographs or images that require a very large colour palette, such as those carrying intricate shading detail.

Bitmap graphics are made up of individual pixels, each of which can be coloured separately. If a bitmap image is resized, the quality of the image is lost and sharp lines become fuzzy and blurred. Bitmap file sizes are generally larger than those of vector images.

Skills check ⟩⟩

Look at page 126 in the skills section on artwork and imaging for more information on bitmap graphics.

Bitmap images can be developed in a similar way to vector images – that is, in stages, building up to a completed picture. Basic shapes are available, and you can also use freehand tools.

Drawing and painting bitmap images

This section shows how a simple drawing of a CD player can be created using draw, paint and freehand tools.

Figure 3.35(a) shows the oval outline of the player and one of the buttons. Both are filled with colour and effects using the **Color Fill** tool 🖾. The track window was created using the **Basic Shapes** tools, filled with colour and effects, as shown in Figure 3.35(b). The **Text** tool **A** was used to write the wording over the top. The earphones were drawn using the **Pencil** tool 🖉 and then filled with colour as shown in Figure 3.35(c). The design was gradually built up until the CD player was complete, as shown in Figure 3.35(d).

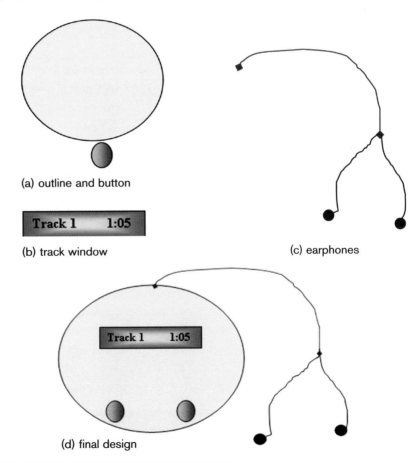

(a) outline and button

Track 1   1:05

(b) track window

(c) earphones

(d) final design

Figure 3.35 The CD player created as a bitmap

Editing bitmap images

Bitmap images can be edited in a similar way to vector images. Use the **Select** tool 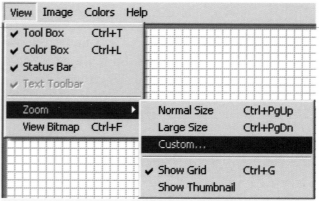 and then cut, crop or move the selected area. Bitmaps can also be edited more precisely at individual pixel level. The individual pixels become clearly visible by zooming to 800 per cent using the **Custom** option and displaying the gridlines (Figure 3.36).

Figure 3.36 Using the zoom facility and displaying gridlines

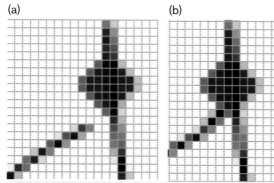

(a)   (b)

Figure 3.37 (a) There is a piece of wire missing.
(b) The missing pixels have been added

If you look carefully at the completed image of the CD player, you will see a slight gap in the wire for the earphones just where the two wires join together. When the image is zoomed, this gap becomes very clear, as shown in Figure 3.37(a). Using the pencil and selecting the black or grey colour, the missing pixels can be filled in to complete the wire, as shown in Figure 3.37(b). If you need to delete a small section of the bitmap, the **Eraser** tool provides very fine control with its four different sizes.

Adding text to bitmap images

Text can be incorporated into a document by clicking on the **Text** tool **A** and placing the cursor where you want to enter the text. The text will be an object that can be moved around. Applications such as Corel Photo-Paint provide a variety of text effects similar to those available in vector formats: 2D, 3D, shadow, etc.

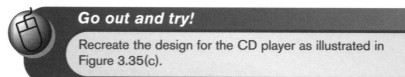

Go out and try!

Recreate the design for the CD player as illustrated in Figure 3.35(c).

Open your file 'Unit 3'. Create a new bold heading **'Bitmap images'** and write a paragraph describing the skills used to design the CD player. Save the file.

Combining bitmap images

Once several objects are exactly the way you want them, they can be combined into one object. After doing this, there is no risk of accidentally moving part of the image you have created, and the combined image can be resized as a whole.

Skills Builder Task 4

At this stage you should be able to tackle Task 4 of the Skills Builder mini project on page 56. Don't forget to study the scenario carefully so that you are clear about the project objective.

Applying special effects to bitmap images

There is a wide range of special effects that may be applied to bitmap images, including sharpen, soften, negative, emboss, watercolour, stained glass, chalk and charcoal. Some are illustrated in Figure 3.38.

original picture of roses

stained glass effect

negative effect

chalk and charcoal effect

Figure 3.38 Special effects added to an image of roses

Sometimes, applying one of these effects is not successful; at other times, it can make a definite impact. You just have to try out an effect to decide whether it achieves your aim. For example, the stained glass effect on the roses could possibly be used if you wanted to illustrate it as if it were a jigsaw puzzle.

Go out and try!

1 Select a suitable photograph from a clip art library and apply effects such as stained glass, negative, chalk and charcoal, or any others of your own choice.
2 Working in groups of two or three, compare the images and discuss which effects you think are the most effective and in what circumstances they might be useful.

Open your file 'Unit 3'. Create a new bold heading **'Special effects in bitmap images'** and write a paragraph describing the skills you have learned. 💾 Save the file.

Skills Builder Task 5

At this stage you should be able to tackle Task 5 of the Skills Builder mini project on page 56. Don't forget to study the scenario carefully so that you are clear about the project objective.

Using layers in bitmap and vector images

Graphic images typically consist of two or more individual items that together have made the completed image. Frequently these individual items are placed one on top of the other and then grouped together. If you need to edit a design that has been grouped, you then have to ungroup it in order to make the changes to the relevant object.

However, if you use *layers*, each layer is a separate image that can be edited on its own. Think of a layer as an image on a sheet of clear material. Together, all the layers form a stack of images, but the order can be changed as necessary.

Figure 3.39 Layers as they appear at the side of the screen

The picture of the dogs playing in the snow (Figure 3.39) was built up in stages (there are many image manipulation programs that will provide these facilities). Some graphics software allows you to combine vector and bitmap images into the one document. In this case the photographs are bitmap images, but the text is a vector image. Edits were made to the various layers, and layers were moved up and down the stack.

- In stage 1, the photograph of the garden was opened.
- In stage 2, a suitable photograph including a dog was found on the Internet.
- In stage 3, the outline of the dog was cropped, copied, flipped horizontally and positioned in the garden scene.
- In stage 4, the text was written. It was later changed to 'What's this stuff?'

Figure 3.40 shows the final layers for the completed graphic.

Figure 3.40 Final layers for a completed graphic

Open your file 'Unit 3'. Create a new bold heading '**Layers**' and write a paragraph describing the skills you have learned.
Save the file.

Skills Builder Task 6

At this stage you should be able to tackle Task 6 of the Skills Builder mini project on page 56. Don't forget to study the scenario carefully so that you are clear about the project objective.

Preparing images for screen publication

In preparing images for screen or web publication you will need to consider

- the qualities of the graphic
- colours and text (fonts) suitable for on-screen viewing
- the file format
- the size of the file.

Considering the qualities of the graphic

Image resolution

The typical computer monitor has 72 or 96 pixels per inch (ppi). In other words, there are 72 or 96 dots of colour in each one inch (2.54 cm) horizontally and vertically.

Suppose you have an image that is 400 pixels wide and 400 pixels tall. If you need to display this in a two-inch-square box on the screen, the drawn image will have an effective resolution of 200 ppi

Skills check

Look at page 126 in the skills section on artwork and imaging for more information on image resolution.

(200 pixels spread across each inch). Since the monitor's resolution is only 96 ppi, not all of the pixels can be drawn. To compensate the computer is effectively displaying approximately alternate pixels, which means that some detail will be lost.

As some detail will be lost anyway, you could resize the image using graphic software to make the file smaller. The image will then load more quickly and take up less memory.

large image

both images look identical when fitted to a small box on screen

resized image

Figure 3.41 An image for on-screen viewing does not need a high resolution

Display monitor characteristics

The maximum resolution available to display images on a screen depends also on the type and screen size of the monitor. There are many types of monitors available today. Monitors based on cathode-ray tube (CRT) technology now range from 15 inches to 21 inches in size – measured diagonally across the tube face (not all of which is viewable). Most monitors have an aspect ratio of 4:3, which means that the width is larger than the height. A 19-inch (48-cm) CRT monitor actually has a viewable area of approximately 14 inches width and 11 inches height. (Despite the fact that the UK started to adopt the metric system of measurement in 1965, monitor screen sizes are still stated in inches.) The size of a liquid-crystal display (LCD) monitor is measured along the diagonal of the actual viewable area, so an 18-inch LCD monitor will provide a similar viewable area to that of a 19-inch CRT monitor.

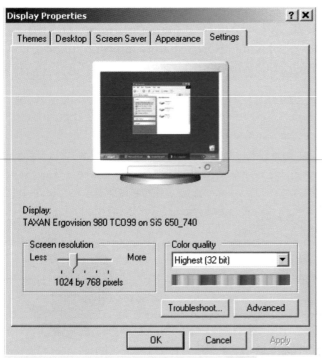

Figure 3.42 Dialogue box for Display Properties

The resolution of a monitor is the number of pixels that can be displayed on the screen, usually given as the number of columns followed by the number of rows. Currently 800×600, 1024×768 and 1280×1024 are the most common display resolutions. Since about the year 2000, 1024×768 has been the standard resolution. Many web sites and multimedia products are designed for this resolution.

Figure 3.42 was produced on a 19-inch CRT monitor with a resolution set at 1024×768 pixels. You will notice that this is approximately a third of the way along the scale for the screen resolution; on a 14-inch monitor, this would probably be at the maximum. The resolution of the 19-inch monitor used can be increased to 1600×1200 pixels. The advantage of the higher resolution is that you can display large pictures without having to zoom in to see the detail.

Many people tend to think that the picture on the screen is static. This is true for an LCD monitor, but on a CRT monitor the picture is continually 'redrawn'.

A refresh rate of 60 Hz (hertz) allows the screen to refresh itself 60 times per second. If the refresh rate is too slow then the picture will no longer appear static but will flicker and be uncomfortable for the viewer, possibly producing eyestrain.

Jargon buster

The **refresh rate** is the speed at which a CRT monitor screen image refreshes or 'redraws' itself.

TiP

The ideal refresh rate is set for the monitor and should not be adjusted.

Go out and try!

1 Right-click with your mouse on the desktop area on the screen. The *Display Properties* dialogue box will appear (Figure 3.42).
2 Select the *Settings* tab and look at the resolution of the monitor you are using.
3 Make a note of the pixel setting now in case you need to revert to it after changing it.
4 Change the resolution to the maximum allowed by the scale (drag the pointer to the right), and click **Apply** to find out the effect this change has on the desktop icons.
5 You will be given the option to retain the settings – select **No**. If you do lose the settings, you can repeat the process to revert back.

Open your file 'Unit 3'. Create a new bold heading '**Screen resolution**' and write a paragraph describing the skills you have learned. Save the file.

Choosing colours and text suitable for on-screen viewing

Figure 3.43 Color mode options

When creating images, you have a choice of colour systems to use, as shown in the *Color mode* text box (Figure 3.43). The option you choose will depend on how you want the finished graphic to look, and considerations of file size.

What traditionally was called 'black and white' is in graphics terminology known as *greyscale*. The 'black and white' option means exactly that – there are no shades of grey – and it is used for cartoons or any other line drawing that does not involve colour or tints. Figure 3.44 shows the difference.

✔ TiP

If the image is in black and white, in greyscale, or is not going to be printed in colour, there is no point in preparing the artwork in full colour. The file size will be much larger for full colour images compared to greyscale or black and white.

Figure 3.44 (a) Sydney Opera House in black and white. (b) The same image in greyscale

✔ TiP

The term **colour system** can also mean the way colours are encoded – see the definitions of **RGB** and **HSB** on page 47.

Go out and try!

1 Find a clip art image of your choice and save it in full colour.
2 Convert the image to greyscale. To do this in Corel Photo-Paint, select **Image, Convert To** and then **Grayscale**. Save the file as 'Greyscale'.
3 Convert the image to black and white. Save it as 'Black and white'.
4 Compare the original and the two new images. Which one do you prefer? Which one do you think is the most effective?
5 What are the three file sizes?

Open your file 'Unit 3'. Create a new bold heading '**Using different colour systems**' and write a paragraph describing the skills you have learned. Save the file.

Colours suitable for the web

The use of so-called 'web-safe' standard colours may not be so crucial now that more and more computers are able to display millions of different colours. When you design a website, it can be very tempting to use a wide range of colours, but it is essential that you consider your audience, rather than just your personal preferences.

We do not all see colours the same, and some people are *colour blind*. That does not necessarily mean that they don't see *any* colours, but a typical problem can be that very dark colours all appear black, or orange on a red background may not be visible.

Fonts suitable for the web

The fonts you use for on-screen viewing should not be too small or too fancy. Of course you will use interesting styles for your text, rather than plain styles that are appropriate for reports and business documents, but make sure the text is clear and easy to read.

Go out and try!

1 Visit the websites W3 Schools and BT Age & Disability Action (via www.heinemann.co.uk/hotlinks (express code 9842P)) to see the range of 'web-safe colours' available and to learn more about colour-deficient vision.

2 Study the two rectangular blocks in Figure 3.45. How many colours can you see in each?

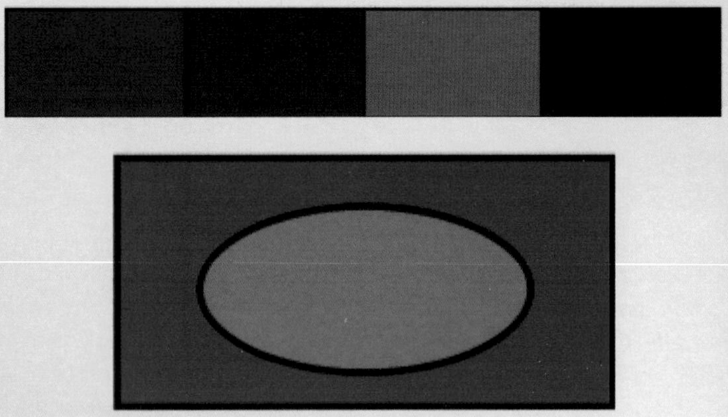

Figure 3.45 Try to see the colours in these filled rectangles

Open your file 'Unit 3'. Create a new bold heading '**Colours and fonts for the web**' and write a paragraph describing what you have learned. Save the file.

Colour balance

If the colour of an image is not effective, you can use the **Colour Balance** tool to adjust it. For example, the photograph of a windmill shown in Figure 3.46(a) has a very dull sky. By adjusting the 'red' colour balance from 0 to –23, we made the sky more blue and the whole picture noticeably brighter.

Figure 3.46 Windmill photo before and after improving the colour balance

Levels and curves

If you wish to change the brightness or contrast of an image in software such as Microsoft Photo Editor, there is a simple control to make these adjustments. However, any changes will affect the whole image. For example, if a photograph is too dark in just one area, adjusting the contrast will over-expose the areas that are correctly exposed, making these sections too light: You will have corrected one problem, but created another one.

More sophisticated software, such as Adobe Photoshop, provides tools that enable you to make more controlled changes.

1. The **Levels** tool enables you to adjust individually three aspects of an image: the shadows, midtones and highlights. These three aspects can be adjusted for part of the image or for the whole image.

2. The **Curves** tool provides even more versatile control for making changes to an image. You can adjust up to 14 different points throughout an image's tonal range, from shadows to highlights.

The use of the levels and curves tools is very complicated, and requires a great deal of skill and experience to obtain effective results. Likely users would be professional photographers, magazine editors and serious photographic enthusiasts.

Rendering text as an image

Normal text is entered into a document and may be edited by highlighting it, increasing the font size or style, or perhaps changing it to bold or italics. The text cannot be 'lifted' and moved around the page in the same way as an image.

However, when you create text in WordArt, for example, the text is an 'object' which can be edited just by dragging its handles, and it can be 'lifted' and moved around the page.

Sometimes you may need to change standard text into an object. Programs such as Flash enable you to do this. Let's imagine you wish to create a link to a spreadsheet file and you have created a text box with the words 'Link to spreadsheet'. Although you can create a link to the text, there is a disadvantage to the user in that it can be tricky pointing to the right spot to activate the link. However, if you select the box and, from the **Image** menu (Figure 3.47) select **Convert to Symbol** and then **Button**, the text will look the same but will become an object. The object can be positioned more flexibly on the web page and the user can click anywhere in the object to activate the link.

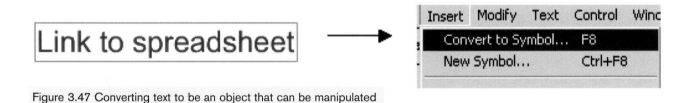

Figure 3.47 Converting text to be an object that can be manipulated

Skills check ⟫

Look at pages 67 and 129 for more information on file formats.

Considering the file format of an image

There are numerous file formats for graphic images. The file sizes vary considerably, but one thing is constant – the larger the file size, the slower it will download for viewing on a user's system. The table in Figure 3.48 highlights the key points for file formats.

File format	Features	Advantages	Disadvantages
BMP	Traditionally used for Windows Colour quality as good as TIFF		Cannot be compressed Files very large
GIF	Supports only 256 colours Excellent quality where there are only a few distinct colours or sharp contrasts Popular for storing stand-alone animated files	Files can be compressed even more than JPEG Ideal for fast electronic transfer of data Excellent choice for online publishing of photographic images	May be subject to royalties
JPEG or JPG	Most suitable for photographs or scanned images Full colour or greyscale	Files can be considerably reduced or compressed Excellent choice for online publishing of photographic images	Not so effective for text, cartoons or black and white drawings May lose clarity if the file is compressed to a very low quality
PDF	Entire content and layout of documents are retained when transferred into PDF format	Can be viewed on any computer system using Adobe Acrobat Reader software, which is free Ideal for sending files via the Internet	If you want to edit the files, not simply to look at the page, then you will have to pay for the software.
PNG	Supports millions of colours	Better compression than GIF Not liable to royalties	Not ideal for printed graphics Does not support animations
PSD	Native file format for Adobe PhotoShop	Facilitates advanced editing	File size tends to be large
SWF	File format for Flash Vector-based format for animations on the web	Very fast download times Images can be resized without affecting quality	Will need Flash player to play back the files – comes as a stand-alone program or a plug-in that 'snaps' into your browser
TIFF	Supports 16.7 million colours Can be any resolution and greyscale or full colour	Flexible Files can be compressed Produces excellent printing results	
WMF	Used to store most Microsoft clip art files and for transfer via the clipboard Suitable for vector and bitmap graphics	Windows-based file format	Some other operating systems can support this format, but not all

Figure 3.48 Features, advantages and disadvantages of file formats

TiP

Compressing a file to a small extent (by about 20 per cent) can achieve a big reduction in the file size (by about 80 per cent) without impacting on quality.

 TiP

FrontPage will automatically convert to GIF format any image files that you insert into your web page, if they are not already in GIF or JPEG format.

Compression and compressibility

Ideally you want the images you use on a website to look just as they did on your screen in the original file format, but you also want the file sizes to be small so that download time is short. The standard compressed format is JPEG, which reduces the file size by removing data. Some of the image quality will be lost, but as long as the compression factor is not too great this will not be noticeable to the human eye.

The compressibility of a file has reached its maximum level when the reduction in file size has removed too much data, so that the deterioration in the quality of the image is clearly visible. Look at the three photographs of the chair in Figure 79 on page 128. There is a noticeable difference in the quality of the picture for the third chair because the file was compressed almost to the maximum level.

Converting files to different formats

To convert an image to GIF format, the colour mode of the image must be 8-bit (256 colours) or less. To check the colour mode, select **Image, Image Info** and a dialogue box will give the file details (Figure 3.49).

If necessary, convert the colour mode by selecting **Image, Convert To** and choose **Paletted (8-bit)**, as in Figure 3.50. To convert an image to JPEG format, the colour mode should be 24-bit RGB.

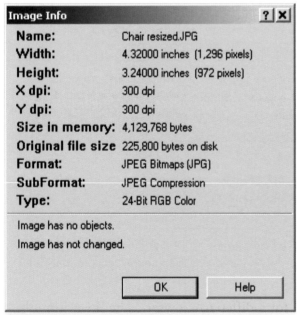

Figure 3.49
You can check the colour mode through Image Info

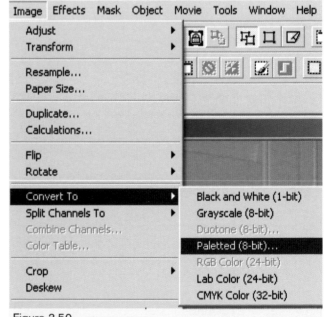

Figure 3.50
Converting the colour mode to 8-bit format

TiP

Websites such as GifCruncher (see www.heinemann.co.uk/hotlinks (express code 9842P) for more) will automatically reduce your GIF pictures without sacrificing quality.

Jargon buster

Transmission speeds are measured in **kbps** (kilobits (1000 bits) per second) or **Mbps** (megabits (1 000 000 bits) per second).

TiP

While it can be tempting to put every possible feature into a web page, it is better to be sparing in your use of special effects, animations, scrolling text and images; this will optimise the site for fast download.

TiP

To make sure your website will be available to all users, keep to the original HTML specification issued by the W3C (WORLD WIDE WEB CONSORTIUM) and you won't have a problem.

Considering further effects of an image file's size

The impact of file sizes has been mentioned numerous times in this book – the larger the file size, the slower it is to appear on the user's screen. We have looked at ways of reducing the file size and considered the impact of

- combining objects
- colour systems
- file formats
- resizing photographs
- compression.

Sometimes, how images are seen on a user's system is out of the designer's absolute control.

Download speed

The speed at which images can be downloaded depends on the Internet service the user has and the quality of his or her connection. The Internet connection may be dial-up or broadband.

The speed of a broadband connection varies depending on the speed of the *modem* used. A 512-kbps modem is up to ten times faster than dial-up, but a 2-Mbps modem is up to 40 times faster than dial-up. With dial-up you pay for the cost of the phone call while you are online, but you cannot then receive or make telephone calls at the same time. With broadband you pay a monthly charge and it does not interfere with your phone calls.

We have become so used to systems with fast reaction times that we can become frustrated if the download speed is unduly slow. If, for example, someone is researching to buy a new car and has to wait more than about 30 seconds for an image to download, he or she is likely to switch to a different website.

Browser characteristics

The two most commonly used Internet browsers are Microsoft Internet Explorer and Netscape Navigator. Both have retained the core HTML coding, but have added their own personal touches – their own HTML tags. They also use different methods to render websites on screen.

If a website was created for a different browser technology from the one on your system, you may not be able to view that site or use all of its features.

Skills Builder Task 7

At this stage you should be able to tackle Task 7 of the Skills Builder mini project on page 56. Don't forget to study the scenario carefully so that you are clear about the project objective.

Preparing images for print publication

When you prepare images for the screen, you can see what they will look like, but when you prepare images for other mediums – paper, acetate or fabric – you can't guarantee that the colours will look the same. For example, you may have bought clothes by mail order that turned out to have different colours from those you had been expecting, because of the way the photos appeared in the printed advert.

You also need to take into account other factors, such as

- printer characteristics
- the medium
- layout considerations
- viewing distance
- colour management.

Printer characteristics

Types of printer

There are two main types of printer that you are likely to encounter at home or in school or college – *laser* and *inkjet* (also known as *bubblejet*).

- Laser printers may print in black and white or in colour. They are quiet and are used for high-quality low- or high-volume work.
- Inkjet printers have one or more heads carrying coloured inks, with the best ones providing a wide range of colours. They are rather slow but generally cheaper than laser printers and are suitable for all types of high-quality low-volume work.

Laser printers produce better definition for text and graphics, but inkjets produce superior results for printing photos. The speed and quality of print will depend on the specification of the printer, which is often related to the cost of purchase.

Resolution and other preferences

When you are printing primarily text documents, it is not usually necessary to change the default settings for the printer. However, for the printing of graphics, printers provide 'printing preferences' – different resolution options – to improve the quality of the image.

Figure 3.51 shows the graphics options for a black and white laser printer. Notice that the 'Best' resolution is 1200 dpi (dots per inch), double that of the 'Normal' setting. The advanced options also allow you to change the 'Darkness' of the printout.

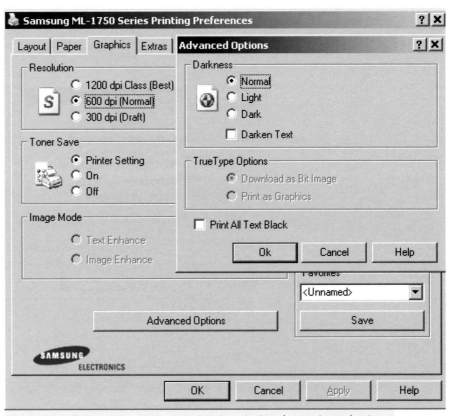

Figure 3.51 The graphic options of a black and white (monochrome) printer

---

TiP

Printers often have a 'draft' option which produces a lower-quality output. This is satisfactory if you wish to save ink and are not printing the finished version of your work.

A colour printer will generally provide a wider range of printing preferences for the user to choose from. Figure 3.52 indicates that the default setting for this inkjet printer is 'Text'. Notice that in addition to draft, you can also select from 'Text & Image', 'Photo' and 'Best Photo'. The latter will be the highest resolution and best quality. In addition there is a wide variety of paper options, depending on whether you are printing on paper, photo paper or transparencies (acetate). For this inkjet machine, selecting the 'Advanced' option produces a warning not to make changes unless you are an experienced user.

Figure 3.52 Printing preferences of an inkjet colour printer

The medium

As already mentioned, when preparing images for print, the medium on which you are printing may be plain paper, photographic paper, card, acetate or fabric. You are unlikely to print on fabric yourself, but remember to check the printer settings for the right medium, as explained above.

Layout considerations

Paper size and orientation

Study Figure 3.53 for a moment. It shows the standard sizes of paper used in many countries (not the USA).

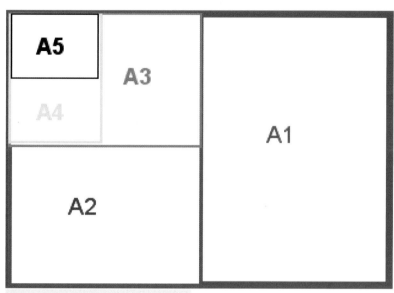

Figure 3.53 Standard paper sizes

A0 is the largest of the standard sizes for paper and has an area of exactly one square metre. Then

- A1 is half of A0
- A2 is half of A1
- A3 is half of A2
- A4 is half of A3
- A5 is half of A4.

A0 and A1 are very large and are used for technical drawings and posters. A1 and A2 are the common sizes for flip charts. A3 is used for drawings and smaller posters. A4 (210 mm by 297 mm) is the paper size that you will be most familiar with, and you will almost certainly use it for printing your work. Sometimes you may use A4 paper folded in half to make an A5 leaflet, or folded in three to make a leaflet with three columns on the inside of the fold. Postcards are half the size of A5 and – yes! – they are known as A6.

Every size of paper can be used in portrait or landscape *orientation*.

Margins, gutters and bleeds

You will need to consider whether to change the default settings for the margins. The term 'gutter' refers to a margin setting that adds extra space to the side or top margin of a document that will be bound (such as a book) to prevent the binding obscuring the text. 'Bleed' allows printing right to the edge of the paper. It is ideal for posters and photos, provided you have a printer with this capability.

Viewing distance

When you are looking at your work on the monitor, you are obviously seeing it at close quarters. If the size of the text and images is clear on screen, your document will probably be fine for a leaflet. However, if you are designing a poster that will generally be viewed from a distance, it is important to ensure that the words and images are much larger.

Colour management

If you send a colour image from the computer to two different printers, even if they are exactly the same model, it is quite likely that the results will be different. This is not a problem for personal use, as long as the results are satisfactory. However, for commercial use this is a potential problem. If a manufacturer of wallpaper or fabrics sends out samples to customers, it is crucial that the colours be accurate. Printer calibration tools will ensure that each digital device – monitor, scanner, printer – is standardised to produce consistent results. You are not likely to be able to implement calibration procedures, but it is important to be aware of when it would be necessary.

Skills Builder Tasks 8, 9 and 10

At this stage you should be able to tackle Tasks 8, 9 and 10 of the Skills Builder mini project that follows. Don't forget to study the scenario carefully so that you are clear about the project objective.

Skills builder mini project

Skills Builder mini task

This activity provides an opportunity to develop the skills you will need in order to complete a summative project for Unit 3. Your teacher or tutor may decide to use the tasks in this mini project as further practice as you work through the unit, or may alternatively set it as a mini project all at once. If it is set as a mini project then, before you begin, you should think about how to plan and manage the project – refer to page 144.

The scenario

Your school/college has recently completed a major rebuilding programme and, to complete the transformation, needs an updated corporate identity. In this respect the governors have decided that initially they would like a new logo, updated stationery and a new brochure. With particular regard to the brochure, you have been asked to produce a design for the front and back cover, together with a double-page spread that will illustrate specific design features of the brochure, such as

- *a background design/style layout for the page*
- *graphic styles for main headings and sub-headings*
- *icons to represent specific course types.*

In addition, there is to be a design feature on every page to include passport-style photographs of students and short comments from them about their courses – similar to the features identifying tips, jargon busters and so on in this book.

The Principal knows that you have developed a wide range of graphic skills and has asked you to help with this project.

Task 1

Use secondary sources to put together a collection of relevant stimulus materials by investigating a range of

- corporate logos
- brochures
- icons and other design features.

Make sure that each item in your collection is catalogued and acknowledged appropriately.

Task 2

Use primary sources to capture a range of images that will be appropriate for inclusion in your brochure. Ensure that each item in your collection is catalogued and acknowledged appropriately.

Task 3

Use vector-based software to produce *three* logo designs for your school/college, exploring a wide range of styles and techniques. Test your designs by showing them to at least three of your friends and three adults. Record their comments and, taking these comments into account, produce one final product.

Task 4

Use bitmap-based software to produce icons designed to represent the different courses on offer, exploring a wide range of styles and techniques. Test your designs by showing them to at least three people. Record their comments and, taking these comments into account, produce your final products.

Task 5

Design a double-page spread of the inside pages of the brochure, incorporating examples of the design features outlined in this project.

Task 6

Think about the front and back covers, and include a digital photograph of your school/college for the front and a map on the back cover. You could practise layering techniques to add in the logo, suitable text and any other relevant topics to enhance your design.

Show your design to at least three people and record their comments. Edit the cover, taking into account these comments, and produce the final product.

Task 7

Design a single web page to advertise your college. Include the college logo, a photograph or video clip, map and some supporting text.

Task 8

Design a poster, incorporating the logo and a photograph plus any other suitable material, to advertise an open event for prospective new students, identifying the factors to be considered when printing a poster compared to printing on a typical A4 sheet.

Task 9

When you have finished your project, review your work and produce a short evaluation outlining how well it met the requirements explained in the scenario and any aspects you feel could be improved. Justify your comments. For example, if you say the design layout is effective, explain what makes it effective. If something still needs changing, explain why and how.

Assessment Hint

You should attach to your evaluation the records of the feedback you received and the actions you took in response.

Task 10

Finally, present your work in an e-portfolio using suitable file formats. Your e-portfolio should be designed to present the following:

- home page
- table of contents
- the finished logo
- a selection of icons representing college courses
- the design of your college brochure, including front and back covers and the double-page spread
- a new web page.

You should also include some supporting evidence:

- your project plan
- the stimulus material you collected, showing full acknowledgement of the sources
- evidence of the development of the design of your logo, including details of feedback from other people
- evidence of the development of the design of the icons, including details of feedback from other people
- evidence of the development of the design of the front and back covers and the double-page spread
- evidence of the development of your web page
- review and evaluation of the project, including feedback from others.

TiP

Throughout your project, in order to achieve top marks you must

- *create appropriate folder structures*
- *use suitable filenames and file formats*
- *carry out regular backup procedures.*

You will find that the requirements for the e-portfolio, project planning and review and evaluation are almost identical for each of the units. If you follow the guidelines given in the relevant chapters in the book it will help you to achieve top marks.

PART 2

Section 1 | Standard ways of working

While working on your projects you will be expected to use information communication technology (ICT) efficiently, legally and safely. So it is important to understand the need for good practice both in your studies and your future work. The guidelines in this section will ensure that you work with ICT in the correct manner and avoid problems that may cause you to lose your coursework at a crucial stage.

LEARNING OUTCOMES

You need to learn about

✓ file management

✓ personal effectiveness

✓ quality assurance

✓ legislation and codes of practice

✓ working safely.

File management

You may be absolutely brilliant at creating artwork and producing imaginative websites, but is your file management system just as brilliant? If it becomes so disorganised that you cannot remember what a file was called, waste time trying to find the right file, or have no backups, you may not achieve your qualification simply because you were unable to hand in the assignment!

In a work environment, your employer would become frustrated with you if, despite being very capable, you never met your deadlines because of poor file management skills.

So what do you need to do?

- Save work regularly and make backups.
- Use sensible filenames that indicate their contents.
- Use appropriate file formats.
- Set up folder structures.
- Limit access to confidential and sensitive files.
- Use effective protection against viruses.
- Use 'Readme' files where appropriate.

Saving work regularly and making backups

Probably the most obvious and simplest rule to remember is to *save* your work regularly. It is so easy to do, but just as easy to ignore!

You might be so busy designing a superb poster or flyer, thoroughly enjoying working with the graphics and using your imagination, that you forget to save the document. Then, when it is almost finished, you lose *all* the work because there is a power failure or you accidentally delete everything. If you are lucky, when you reopen the application software the file will be 'recovered' by the computer, but you can't guarantee this.

As well as saving your work regularly, it is important to keep backups. Information stored on a computer can be corrupted or lost through a power surge or failure, or by damage to or failure of the hardware. If 'disaster' strikes then you can use the backup copy. It might be slightly out of date, but at least it can be updated.

In business, backups are usually made at the end of every day. At the end of the week another backup is made, and stored separately from the daily backups. To avoid the damage done by viruses or other system-wide disasters, it is not enough to back up files into a different folder on the same computer – files should be regularly backed up onto external storage media. Here are some examples.

- *Floppy disks* are used for fairly small amounts of data – up to 1.4 megabytes (MB). However, files containing graphics quickly become too large to save on a floppy disk.
- *Pen drives*, *flash drives*, *keychain drives* and *memory sticks* (Figure 1) are portable devices that connect to a USB (*universal*

TiP

*As soon as you start a new file, save it with a sensible name. After that you can just keep clicking the **Save** icon to update the saved version.*

TiP

It is a good idea to keep separate copies of your files each time you make major changes. This way you can record your progress in your e-portfolio. Also, sometimes you decide that an earlier version was better after all, so you can easily go back to it.

Figure 1 Memory stick

Jargon buster

One gigabyte (1 GB) equals **approximately** 1000 megabytes. One terabyte (1 TB) equals **approximately** 1000 gigabytes. Text of 500–600 pages uses about 1 MB of space.

serial bus) port of a computer. They can store between 32 MB and 2 GB (gigabytes) of data. These removable drives are the latest method of data storage and are automatically detected by the computer. You can use them to store or move files that won't fit on a floppy disk. These drives are extremely small so they are ideal for transferring files between work or college and home.

- A second hard drive can be installed, either within the same computer or externally. The disadvantage of a second internal drive is that it cannot be stored off-site or in a fireproof safe.

- *Zip drives* are another magnetic disk medium, with storage capacities of up to 250 MB. These are also suitable for domestic or small business use.

- *CD–R* (recordable) and *CD–RW* (re-writable) compact discs can store up to 700 MB. Most modern home PCs are now fitted with built-in CD–RW drives, making this a convenient and inexpensive method of backing up large amounts of data (such as the text and illustrations for this book).

- *DVDs* look the same as CDs, but can store up to 9 GB of data. Several different types of recordable DVD are available.

- A *tape streamer* is a magnetic tape generally used by large businesses. Cartridges that can store up to 300 GB of data are currently available. This method is the most affordable method of backing up the extremely large amounts of data required by large businesses. It also has the benefit that the tapes are small and, as they are removable, they can be stored off-site or in a fireproof safe. The disadvantage of making backup copies on magnetic tapes is that to recover a particular file you must search through the tape starting from the beginning until you reach the file you want – just like with a tape for music.

Using sensible filenames

When working on any project, *as soon as you have written just a few words on the page*, save the file – naming it in such a way that it is easy to identify later. By having a sensible system for all your filenames, it is easy to find the right one without having to search through (and possibly open) several files.

Older operating systems limited filenames to eight standard characters, but modern operating systems allow more than enough space to identify each file clearly.

With Microsoft operating systems, files usually have more characters, called *extensions*, added to the filename. Filenames are followed by a stop (.) and then an extension of up to three characters. When you save a file, the file extension is usually added automatically to the name you have given the file.

Why is an extension necessary? The extension tells the computer which program to start so that the file can be opened. For example, a Microsoft Word (word processor) document will be followed by **.doc**, an Excel (spreadsheet) file by **.xls**, and a bitmap graphic by **.bmp**. Another important extension is **.exe**, which means an *executable program* – an application. Figure 2 is a list of the most common file extensions.

Extension	Application or type of file
.bmp	Microsoft Paint bitmap image
.doc	Microsoft Word document
.dot	Microsoft Word template
.exe	Executable program
.gif	Graphics interchange format image
.html	Hypertext markup language (web page)
.jpg	Joint Photographic Experts Group (JPEG) image
.mdb	Microsoft Access database
.pdf	Adobe portable document format
.ppt	Microsoft PowerPoint presentation
.pub	Microsoft Publisher publication
.swf	Macromedia Flash vector graphics
.tiff	Tagged image file format image
.tmp	Temporary file
.xls	Microsoft Excel spreadsheet
.xlt	Microsoft Excel template

Figure 2 The most common file extensions

Choosing appropriate file formats

In most cases the file format is automatically selected by the software application. For example, Excel will always use the extension **.xls** for spreadsheets and **.xlt** for templates.

When saving images you often have a choice about which format to use. The screen shots for this book were saved in Corel PhotoPaint and the *TIFF Bitmap* format was chosen (Figure 3) as it allows 16.7 million colours. TIFF graphics are very flexible: they can be any resolution, and can be black and white, greyscale, or full colour. It is the preferred format for desktop publishing as it produces excellent printing results.

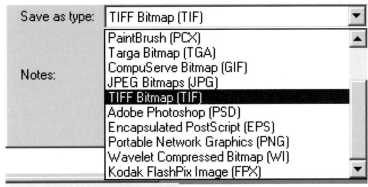

Figure 3 File formats

JPEG format is the most suitable format for full-colour photographs or greyscale images, such as scanned photographs, with large variations in the colour. It is not so effective for text, cartoons or black and white line drawings.

The maximum number of colours a *GIF format* image supports is 256, much less than the TIFF and JPEG formats. However, GIF format is significantly better for images with a just a few distinct colours, where the image has sharp contrasts – such as black next to white, as in cartoons.

PNG (Portable Network Graphics) is an alternative format for a variety of applications. PNG offers many of the advantages of both TIFF and GIF file formats, but the file size is smaller. This may be an important consideration, as your complete e-portfolio will be limited to a maximum file size.

Each project brief for your e-portfolio will specify acceptable file formats. These are likely to be PDF for paper-based publications, JPG or PNG for images, HTML for on-screen publications and SWF

Assessment Hint

*Files in your e-portfolio must be saved by using the formats **.html**, **.pdf** or **.swf**, so that they are suitable for use on any computer.*

✔ TiP

Sometimes you may wish to rename a file – if you do, be sure to add the appropriate file extension to the new name, otherwise Windows might have trouble in opening it.

✔ TiP

*The JPEG format uses **lossy compression**, which means that you get very small files compared to TIFF format. However, they won't exactly match the original images, resulting in reduced quality.*

✔ TiP

The advantage of the PNG file format is that the compression of the file size is 'lossless' with full 24-bit colour, and can be read by a web browser.

(Flash movie) for presentations, although this may change depending on future developments. To convert files such as Word to PDF format, you will require software such as Adobe Acrobat. Files that have not been created in HTML (or web format), can be converted from Word, Excel and PowerPoint by selecting **File**, **Save As** and choosing the **Web Page** format from the *Save as type* box. You can choose other alternative formats through this method and may need to experiment to find the most suitable one.

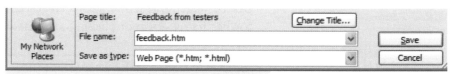

Figure 4 Saving a Word document as a web page

Setting up folders to organise files

Jargon buster

Files relating to one topic can be saved in a **folder** (Microsoft's name for a **directory**). This is an area on the computer's hard disk created in order to organise your computer's file system.

If you save all your files in one place on a hard drive, it can be quite difficult to find the one you want. The solution is to create a main folder for a particular project, and then sub-folders to contain particular elements of the project. Figure 5 shows a main folder called 'Digital Applications', which contains a sub-folder called 'Standard Ways'. All the work for this book was saved in the main folder, and the material for this chapter was saved in the first-level sub-folder. As more and more chapters were written, more sub-folders were created to contain them. One sub-sub-folder is called 'Screen dumps', where all the screen shots were saved in TIFF format.

If you need to copy or delete files or folders, open Explorer as explained opposite, right-click on the relevant file or folder and select **Copy** or **Delete**. If you are copying the file or folder, move your mouse pointer to where you wish to place the copy, right-click the mouse and select **Paste**.

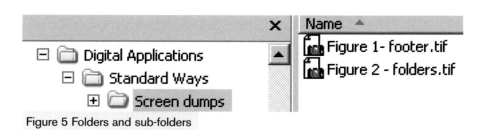

Figure 5 Folders and sub-folders

Go out and try!

1 To make sure you are organised right from the start, create the following folders and sub-folders to save the work you will produce for these chapters:

Standard Ways
ICT Skills
   Word-processing software
   Presentation software
   Artwork and imaging software
   Internet and intranets
Creating an e-portfolio
Project Planning
Review and Evaluation

○ From the **Start** menu, right-click the mouse and select **Explore**.

○ In the left-hand pane, click on the drive where you want to create your folders.

○ The contents of the drive are shown in the right-hand pane. Right-click in the white space of this section. The following menu appears.

TiP

Organising your electronic files may vary, depending on the software available to you.

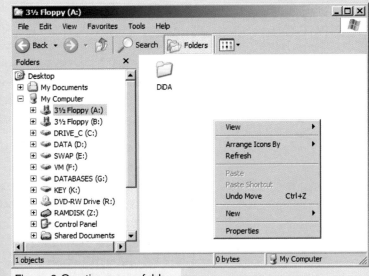

Figure 6 Creating a new folder

○ Select **New** and then **Folder**. The new folder appears.
○ Type in the name for the folder and press **Enter**.
○ Repeat these steps for each of the folders and sub-folders.

Limiting access to confidential or sensitive files

Most files saved on home computers are not particularly confidential or sensitive, but business data held on a computer can often be very confidential or sensitive. For example, medical records should not be accessible to the cleaning staff in a doctor's office or a hospital. Sometimes files in business may be confidential because the company is developing a new product and does not wish its competitors to know its plans.

There are several ways to protect confidentiality.

Using an ID and password to access the computer

Business systems are usually protected by passwords. Users should be prompted to change their passwords regularly. You will notice that when you enter the password, it is displayed as a series of asterisks (**************). This is to hide the characters and prevent anyone reading your password as it is shown on the screen.

Using a password to view data

Modern software systems also allow the use of passwords when saving files. You can set a password to open the file, or a password to modify it. Some people will be able to open the file to *read* it, but they will not be allowed to make any changes.

Many organisations *rank* data files according to their degree of confidentiality. Staff can be given different security levels or privileges which limit access to only *some files* or *some fields* within a file. This is a common method of protecting data and maintaining confidentiality.

Using additional security measures

Internet banking uses a series of security methods to ensure that customers' banking details are kept safe from unauthorised access. Apart from the usual login ID and password, the customer might have to choose another password or some memorable data. This time only certain characters are entered, changing each time the customer accesses the account. This helps to protect against *key loggers*.

For example, suppose the extra password is 'England'. You will see from Figure 7 that on this occasion only the first, fourth and fifth letters of the password are requested from the customer.

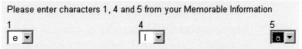

Figure 7 Memorable data to protect confidentiality when using Internet banking

Another security measure is to limit the time that the file or web page can stay open for if it is not used. This ensures that the user's details are not left open on the computer if he or she forgets to close the file. A timeout message will appear (Figure 8).

Figure 8 A time-out warning

Using effective virus protection

TiP

Make sure you run the anti-virus software at least once a week or set it to scan automatically.

A computer virus is a small program that has been developed by someone either for general mischief or to attack a particular organisation. The virus copies itself without the user intending it to, or even being aware of it happening until something goes wrong. Sometimes, in an attempt to prevent detection, the program will mutate (change) slightly each time it is copied.

Viruses can be spread by

- downloading software from the Internet
- opening an attachment to an email
- transferring files from one computer to another via a floppy disk
- using pirated software that is infected.

TiP

It is essential to ensure that the virus definitions are kept up-to-date. Many anti-virus programs will remind you when this is necessary.

Your computer system is far less likely to be at risk if you

- use floppy disks only on one system
- delete emails from unknown sources, especially if they have an attachment
- regularly update your virus protection software.

Viruses can spread very quickly via email. A virus may cause problems such as clearing screens, deleting data or even making the whole computer unusable. Organisations treat the risk of infection very seriously, controlling emails and not allowing staff to take disks between work and home. Anti-virus software is therefore essential, both in business and for the home user.

Anti-virus software

Anti-virus software works by scanning files to discover and remove viruses, a process known as *disinfecting*. There are thousands of viruses waiting to attack your computer. More viruses are being written all the time, so it is essential to keep your anti-virus software up to date. As soon as a new virus appears, the anti-virus companies work to produce a *pattern* file, which tells the software how to discover and stop the virus. Symantec (Norton), McAfee, Dr Solomon's and Panda are some of the companies providing anti-virus software.

Using Readme files

Readme files are files on a CD that give an explanation about how to use the software on that CD. For example, the Readme file might

- explain how to install the software
- describe important points
- give specific information that might apply to your particular setup
- say how much hard disk space is needed to install and run the software effectively
- contain late-breaking information that reveals how reported problems with the software can be solved
- link to additional resources.

Personal effectiveness

To ensure your own personal effectiveness in preparing your project, you must

- select appropriate tools and techniques
- customise settings
- create and use shortcuts
- use available sources of help
- use a plan to organise your work and meet deadlines.

We will look at each of these skills in turn.

Selecting appropriate tools and techniques

If you decide to use video editing in your project, but have access to a video camera and the video editing software for only one hour a week, you are not likely to be able to finish the project. Although it might be a very interesting, exciting project about which you are really enthusiastic, if you can't resolve the problem of access to the equipment it is better to think again and choose an alternative project or presentation style.

Once you have decided on the tools to use, you should then think about suitable, effective techniques that you could apply. For example, if you are fortunate enough to have a video camera and video editing software available, you may need advice on the best techniques to use when making the film. No matter how great your expertise in editing the film, if the original material is poor because there has been too much zooming in and out, and the sweeps across the views were too fast, then the end product is unlikely to be successful.

Customising settings

You can use the Control Panel to personalise your computer in numerous ways, for example by

○ changing the appearance and colour scheme of your screen

○ choosing a screen saver

○ selecting the date, time and language settings

○ changing the mouse buttons for a left-handed person.

Software applications, such as Word, Excel and Access, come with default settings pre-installed during manufacture, but there are many ways to customise the settings to suit your needs. For example, you would probably find the *Standard*, *Formatting* and possibly *Drawing* toolbars to be visible. As you can see from Figure 9, in Word there are 16 toolbars in total – but if most or all of them were shown at once, the screen would be far too cluttered, with not much room left for the page you were working on! Instead,

Figure 9 Toolbars available in Word

(a)

Microsoft
Outlook

Microsoft
Access

Microsoft
Publisher

Microsoft Excel

Microsoft Word

POWERPNT

(b)

Shortcut to
Digital
Applications

Figure 10 (a) Shortcuts on the desktop to software installed on the computer. (b) A shortcut to files for this book

select the ones most useful to you. Word is often set with a default font of Times New Roman 10pt, which is too small for everyday use, so you might wish to set the default to a different size and style.

Creating and using shortcuts

Every time you click on an icon on a toolbar – such as save or print – you are using a *shortcut*. Using the print icon allows you to print the document in a single step. If you use **File**, **Print** and **OK** it is at least three steps – but you might want to choose this method if you want to print two or more copies, or only certain pages.

Each of the toolbars has a list of shortcut icons, but only those most frequently used are automatically visible. You can choose to add any others that you find useful. For example, if you frequently send documents as attachments to an email, you might wish to add the relevant icon to the toolbar.

Shortcuts can also be placed on the desktop. No doubt you are familiar with using shortcuts to software applications – see Figure 10(a). However, you can also choose to create shortcuts to files or folders that you use regularly. For example, a shortcut to the folder containing your work for this qualification might prove useful – see Figure 10(b).

Assessment Hint

Don't be afraid to ask for help, but do remember that the project must be your own work. On the other hand, don't embark on a project that is too complex in case you can't find the help you need.

Getting help

There will be a variety of sources of help available to you, and it is a good idea to take advantage of them. You may find help through

○ software help files
○ textbooks
○ your teacher
○ specialist staff within your school or college

- your peers (who may be more knowledgeable about some aspects of the software, whereas you may know more about other aspects and can offer them help in return)
- family or friends who are experienced in the project you are undertaking.

Using a plan to organise work and meet deadlines

In order to complete your e-portfolio you will have to undertake a number of tasks, gradually building up the content. All the various sections of the e-portfolio will need to be completed to a deadline.

When the final deadline is a long time ahead, it is all too easy to keep thinking there is no rush, and then find you are trying to do ten things at once and cannot possibly finish everything in time! Therefore it is essential to think about all the various items that will go into the e-portfolio and then plan a schedule of your work.

Remember that *it always takes longer than you think*, so make sure you plan to finish well before the final deadline, to allow for all those unexpected delays!

Skills check ▷▷

Information on how to plan and organise your work is presented in more detail in the chapter on project planning (page 144).

Quality assurance

To ensure that your work is accurate and effective, you will need to

- use tools such as spelling and grammar checks, and Print Preview
- proofread your work
- seek the views of others
- check your research.

Using spelling and grammar checks and Print Preview

Spellcheckers

TiP

Be aware that many computer programs use American spellings of words. These are often slightly different from the British spellings.

Spelling errors spoil your work, so always use a *spellchecker* to detect words spelt incorrectly and repeated words (for example, 'and and'). Spellcheckers are available not only in word processors, but also in most other applications such as spreadsheets and email programs. Spellcheckers compare the words you have written against a list in the computer's dictionary, then any words not matching are queried and possible alternatives are suggested.

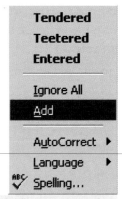

Tendered
Teetered
Entered
I̲gnore All
A̲dd
A̲utoCorrect ▶
Language ▶
ᴬᴮꟲ Spelling...

Figure 11 Adding a word to the spellchecker dictionary

But beware! Although the spellchecker is an excellent tool, it does not understand what you are trying to say and so it can be wrong. This means that *it is still important for the operator to have a reasonable level of spelling!* For example, it will not correct 'whether' for 'weather', or 'to', 'too' or 'two' used in the wrong place.

Sometimes a spellchecker will suggest that a word is incorrect when you know that it is correct. The dictionary will include common names such as 'Smith' but not unusual names such as 'Tenterden', which is a village in Kent. For this the dictionary suggested three alternatives (Figure 11), but the name was spelt correctly.

If you are likely to use a particular word frequently, then it is worth adding it to the dictionary. You can right-click on the word and select the **Add** option. In future the spelling checker will accept the word.

Sometimes words are indicated as incorrect because the language of the spellchecker is set to English (US) not English (UK), so it's worth checking. Select **Tools, Language, Set Language** and choose **English (UK)** and **Default** (Figure 12).

Figure 12 Setting the language that Word will use for its spellchecker

Grammar checker

The *grammar checker* can be more complicated and difficult to use than a spellchecker, and therefore tends to be rather less popular. As a starting point, the best way to use a basic grammar checker is as a tool to draw your attention to possible mistakes. It can be very useful for finding typing errors; such as the one shown in Figure 13, where

Figure 13 The grammar checker showing that there is no space after the bracket

there is no space after the closing bracket. At other times the suggestions may not be clear, so *think about* the suggestions made – the final decision is yours.

Print Preview

Before sending your work to the printer, it is a good idea to check it in Print Preview, where you will often identify poor layout that can be corrected before printing. This saves time and prevents you wasting paper.

When the author of this chapter used Print Preview at this point, it revealed a blank page that hadn't been noticed, which would have interfered with page numbering. It was an easy matter to delete the page before printing the file.

Proofreading your work

Even though you read your work on the monitor screen, it is surprising how often errors can be missed that the spelling or grammar checks have not shown up. In addition, you often find that although you were clear in your mind what you wanted to say, on re-reading the text you realise it is muddled, or you haven't fully explained the topic.

As authors of this book we received the proofs (the first copy of the book as it would look when finally printed) and we had to check that everything was correct, including the page layouts. At this stage we made some changes to our original work, not because it was necessarily wrong, but because it wasn't as clear as we wanted it, or it needed to be a little more detailed.

Seeking the views of others

Once you have an idea about the project you wish to undertake, it is excellent practice to ask other people what they think of the idea.

When writing this book, we discussed the syllabus and decided which of us was going to undertake which sections. We also brainstormed ideas about what should be included, and the presentation. Each one of us came up with ideas and suggestions, and together we produced (we hope!) a better result than if we had worked separately.

We also read each other's material and often asked colleagues to read what we had written. As an author you can be 'too close' to your own project to spot minor errors, whereas someone independent

TiP

If your work is web-based, make sure you check it using different browser software on several computers. Is it always displayed as you had intended?

TiP

Proofreading isn't the most popular task, but it is important. You might discover silly errors, or an explanation that doesn't quite say what you intended. Correcting it before handing in your assignment or coursework might make all the difference to your final grade.

TiP

Do ask someone else's opinion of the work you have produced for your e-portfolio – and do take note of the comments made. A critical evaluation of your project should be viewed as helpful advice to enable you to improve what you have done, not as a negative criticism.

notices odd little mistakes, or a paragraph that is not clear. These can be edited to improve the final version. It is very important to keep draft versions of your work so you can show changes made as a result of comments from others.

Authenticating your work – checking your research

In general, textbooks are likely to contain reliable information because they will have been written to provide students with the information they need for a course they are studying, and are published by reputable companies. If it became obvious that an author or publisher was producing unreliable textbooks, they would soon go out of business: no one would buy their books.

Probably, by the time you read this book, information such as the amount of data a memory stick can hold will have changed. This does not mean that it contains deliberate misinformation – it's just that technology advances. If you do make use of that kind of data given in a book then it would be wise to check whether it is still accurate.

Figure 14 Discussing your work with other people

TiP

In order to authenticate your own work, try researching through **a variety of sources**. If they all match each other, then you can be confident that the information you have obtained is accurate and reliable.

As you are well aware, the Internet is an amazing tool to use when researching almost any topic you could think of. It is, however, all too easy to drop into the trap of *believing everything on the Internet is accurate and reliable*. Unfortunately, that is not the case. Anyone can set up a website and put information on that site. Sometimes the author genuinely believes that the information posted is authentic or accurate, but sometimes the author wishes to deliberately mislead. Therefore, when using websites to obtain information, it is very important to use sites – such as those of the BBC or other well-known organisations – where it is likely that information given will be authentic.

Legislation and codes of practice

To ensure that you comply with legislation and codes of practice relating to the use of ICT, you will need to

- acknowledge your sources
- avoid plagiarism
- respect copyright
- protect confidentiality.

Acknowledging your sources and avoiding plagiarism

In order to complete your portfolio evidence you will need to undertake research. This may be from books, from newspapers or journals, from the Internet, or from people through surveys, questionnaires or face-to-face discussions. It is essential that you acknowledge where you obtained this information – your *sources*.

In some cases you may have to gain permission from the author, artist or relevant organisation to use the material. You should also include a *bibliography* of books, journals, articles and websites that you have found useful in your research.

People may copy original work and present it as their own. This is much easier to do from ICT systems (and especially from the Internet) than it is from a paper-based source such as a textbook. This is called *plagiarism* and, because it breaks copyright law (explained overleaf), it is a serious offence.

It is quite acceptable to *quote* a reasonable amount of someone else's work, as long as you put inverted commas (" ") at the beginning of the text and at the end, and identify the original author, the book, web address or newspaper where you found the information.

Jargon buster

A **bibliography** is a list of sources of information that were useful when undertaking your research, and helped you to form your own opinions when writing an essay or producing the e-portfolio. For example, if you have a project to design a website, you might study other websites first to help you decide what makes an effective site and what doesn't.

The bibliography should include enough information for other people to be able to find the same source you used: the book, journal or newspaper, title of the article, author, date of publication, and web address.

Assessment Hint

*Note that your portfolio evidence will **not** be acceptable if it is a succession of quotes from other people, with very little original effort on your part.*

? **Think it over...**

Check out the sources that we might have used when writing this textbook.

Respecting copyright

You are probably familiar with *copyright* warnings at the beginning of books, rental videos and DVDs, when there is a statement along the lines of *'All rights reserved. No part of this publication may be reproduced or transmitted, in any form or by any means, without the prior permission of the publisher.'* This is also true for most computer programs, published text and images.

Check for the symbol ©, followed by a date and sometimes a name, as shown at the beginning of this book. This indicates that the work is covered by copyright. It is very important to understand what copyright means and to respect copyright law (Figure 15).

- The Copyright Designs and Patents Act 1988 was originally set up in order to protect the work of authors, artists and composers from being reproduced or copied without permission.

- Current European copyright law extends for 70 years after the death of the author/creator. During this period, the work (book, work of art, software, photograph, music score, etc.) may not be reproduced without permission.

- This original Copyright Designs and Patents Act was in existence long before computers were invented. It has subsequently been extended to include computer software, making it illegal to copy applications without permission from the copyright holder.

Figure 15 Facts about the Copyright Designs and Patents Act

When you buy computer software it generally comes with a *licence* that allows you to install, use, access, display or run just one copy of the software on one computer and one notebook. Your school or college will have purchased a network licence in order to be granted permission to run the software on all of its machines.

Imagine a small business that started with just one computer. The owner bought the software for that computer and will have the necessary licence. As the business expands, he might decide to install a network and use the same software on the network, forgetting to buy a new licence. He has in fact broken the law, even though it was not intentional.

This law affects everyone who owns a computer. You must not let other people borrow your software to install on their computers. Similarly, you should not borrow software from a friend and install it on your computer. The software houses have, after all, invested a great deal of time and money in order to develop the software, so you are stealing from them by not paying for the software. It's no different from going into a shop and stealing a box of chocolates or a DVD.

In order to help protect software copyright, the Federation Against Software Theft (FAST) was set up in 1984 to investigate software piracy. The federation will prosecute when instances of illegal copying of software come to their attention. Be warned!

- Passwords should be kept secret. A note stuck to the side of your computer is not a sensible way to remember your password!

- Personal information should not be passed on to third parties at any time. Imagine a situation where an insurance company employee had access to personal information. Think of the consequences if a list of addresses and details of valuable house contents fell into the wrong hands.

- Staff working with confidential data must be very careful not to disclose personal information during a conversation, however innocently.

- Computer screens in public places – such as a doctor's surgery – should not be visible to the general public. It could be very unfortunate if somebody waiting at the reception desk read personal and confidential information about another patient.

Figure 16 Facts about passwords

Protecting confidentiality

The Data Protection Act of 1994, which was updated in 1998, was introduced to deal with the increasing amount of personal data being held on computers and the potential misuse of that personal data. Think of the many different organisations that hold computerised records containing details of our personal lives, school records, employment history, financial and medical records, criminal activities, etc. Under the Data Protection Act, these organisations must ensure that the data remains confidential. They must also allow people to access the data that is being held about them, in return for a small administration fee.

Both employers and employees have a responsibility under the Act to ensure that personal information is not disclosed, however innocently. Staff must be aware of straightforward things they can do to protect the confidentiality of the data. Examples are described in Figure 16.

Skills check

The information on confidential or sensitive files on page 70 is also relevant here.

Apart from personal information, a commercial organisation will want to keep information about the company confidential. They do not want their competitors finding out how well or badly they are doing, and what new products or designs are being prepared.

Organisations have always been at risk from dishonest staff, but the main difference since the arrival of computers is that it is so much easier to obtain the information – you do not even need to be in the room or the building! Inevitably there has to be trust in and reliance on the users of ICT, especially in a business environment. In fact most people are honest and have no intention of defrauding their employer or disclosing confidential information.

It is essential to take security issues seriously, so that you do not *unintentionally* give someone else access to uncensored or private materials. Also, if you *accidentally* discover uncensored or private materials, you must not take advantage of the opportunity, and it may be appropriate to report that a breach of security has occurred.

Naturally we all chat about our jobs, but it is important to know which information should be kept to ourselves, and when it is OK to talk about something to a friend or relative. If in doubt then keep quiet.

Working safely

Employers are responsible for promoting health and safety in the workplace. However, each employee also has the duty to take sensible precautions. For example, if you spend a lot of time using a computer, you may tend to develop backache, eyestrain or repetitive strain injury, but good working practices will greatly reduce the risk of suffering from such problems. It is important to

- ensure that the positioning of hardware, cables and seating is correct
- check that the lighting is appropriate
- take regular breaks
- handle and store media correctly.

Positioning hardware, cables and seating

Desks should have a non-reflective surface and be at the correct height with sufficient space for a computer, mouse mat, telephone and supporting documents. There should be enough space around

Figure 17 The fundamentals of good posture at a computer workstation

the desk for you to change position and vary your movements. It may be beneficial to use a wrist rest to support your wrists.

The ideal working position is to sit with your eyes level with the top of the screen, the small of your back against the chair, and your feet flat on the floor (Figure 17).

Since we all come in different shapes and sizes, it is important to use a chair that can be adjusted to suit the user. A good computer chair will have five feet with castors to increase stability and to allow freedom of movement. You will be able to adjust the height of the seat to suit your own height, as well as the position and tilt of the backrest in order to support your back. If you are short and your feet don't reach the floor, you can ask for a footrest to be provided so that you can support your feet and legs.

Cables should be correctly and safely positioned where no one can fall or trip over them. They should be securely fastened, usually inside trunking that is attached to the walls. They should *never* be trailing across the floor.

Using appropriate lighting

The ideal position for the computer monitor is at right-angles to a window. In order to keep sun off the screen and help reduce glare, blinds should be provided at the windows. If glare is a problem then an anti-glare shield can be fitted. The office should be well lit and the lighting should offer a contrast between the screen and surrounding area. If the light is too dim, documents will be hard to read, leading to eyestrain. If necessary, adjust the brightness of the screen in response to changes in light.

Taking regular breaks

Bad posture is the main cause of backache. If you work *continuously* at a computer monitor you should be allowed to take a short break (say 5–10 minutes) away from the screen after an hour's *uninterrupted* screen or keyboard work. Ideally your job will involve a variety of tasks so that this situation will not arise in the first place.

Handling and storing media correctly

It is important to know how to protect the media you are using, by handling and storing them correctly. Care should be taken when handling floppy disks, CDs and DVDs in order to protect the data. Store them in protective cases to keep them clean, away from dust and moisture. Avoid touching the surfaces, and keep them away from extremes of temperature.

? Think it over ...

How safe is your working environment at home? What could you change to improve it?

Word-processing software

If somebody asked you 'What is word-processing?' you would probably describe it as *the process of creating, storing and editing text-based documents*. While you would be quite right in saying this, in fact word-processing offers so much more.

Word-processing is one of the most effective ways of communicating information to other people through paper-based documents such as letters, reports, leaflets, newsletters, posters and flyers. In fact, effective word-processing skills also underlie successful communication through on-screen publications such as presentations or web pages.

Your word-processing skills are a valuable asset that you will make use of throughout your student and adult life. Word-processing is more than just sitting down at a computer and producing a document without a second thought. It is the skill of designing a document to convey specific information to different groups of people.

LEARNING OUTCOMES

You need to learn about
- ✓ entering, cutting, copying, pasting and moving text
- ✓ formatting text
- ✓ using paragraph formatting features
- ✓ creating, selecting and inserting components
- ✓ using images/objects
- ✓ using spelling and grammar checkers.

Entering, cutting, copying, pasting and moving text

Entering text

Text is keyed in (or *entered*) via the keyboard. When a word will not fit on the end of a line, it is moved on to the next line. This feature is called *word wrap,* and the word processor automatically inserts *soft returns* at the end of each line. These soft returns are adjustable and will move within the document if necessary – such as when you insert an extra word or two in a paragraph or decide to delete some text.

You only need to use the **Enter** key when you want to start a new line, perhaps after a heading or at the start of a new paragraph. The returns that *you* put into a document are called *hard returns* and they are not adjustable by the software. If you display the formatting marks in a document by selecting the **Show/Hide** icon on the *Standard* toolbar, the ¶ character identifies all hard returns.

As you enter text, there are some basic steps you can take to improve the general presentation of the document:

- For most text-based documents, a font size of 11pt or 12pt will be appropriate.
- Remember to leave one clear line space (achieved by pressing the **Enter** key twice) after a heading and between each paragraph. Look at Figure 18, where the ¶ symbol shows that the **Enter** key has been used.
- Nowadays it is common practice when using a word processor to leave only one space following a sentence. However, in the days of the typewriter, it was standard practice to leave two spaces following any punctuation at the end of a sentence. There are many people today who feel that using two spaces between sentences makes your text easier to read because the sentences stand out clearly. You must make your own decision or follow the guidance from your teacher/tutor.
- Leave one space after all other punctuation.
- If possible, use a *fully blocked* style of presentation when producing letters, reports, etc. This means that everything starts at the left-hand margin. Figure 18 is an example of fully blocked working.

Jargon buster

A manual typewriter has a **carriage return key**, which the typist must press between lines to return the carriage to the left of the paper. Because of this the **Enter** key is sometimes called the **Return** key, and the hidden characters it inserts into documents are called **returns**.

Skills check »

Refer to page 88 for more detailed information on font sizes.

TiP

If you study DiDA Unit 4 you will learn how to use styles to add space after headings and between paragraphs automatically.

Travelbug¶
¶
17·London·Road¶
Whychton¶
TO9·3WN¶
¶
Tel: → 543·2134·5678¶
Fax: → 543·2134·5679¶
Email:·travelbug@userve.co.uk¶
¶
¶
¶
There·is·so·much·to·see·and·do·in·Brussels·that·a·weekend·just·won't·be·long·enough!¶
¶
There·are·over·30·museums·to·visit,·fine·restaurants·and·bars,·shops·selling·anything·
from·high·fashion·to·antiques·to·sophisticated·chocolate·creations.··Browse·the·open-
air·markets·or·stroll·around·the·magnificent·cobbled·Grand·Place.¶
¶
A·visit·to·Brussels·isn't·complete·without·a·visit·to·the·famous·Manneken·Pis,·which·
can·be·guaranteed·to·draw·a·crowd.¶
¶
Travel·by·Eurostar·from·Waterloo·and·arrive·in·Brussels·in·2·hours·20·minutes.··
Selected·services·also·available·from·Ashford·International·Station·in·Kent.¶
¶
Prices·start·at·£125.00·for·2·nights.¶
¶

Figure 18 An example of work presented in a 'fully blocked' style

Cut, copy, move and paste

A significant benefit of word processors is that they let you cut, copy, move and paste text. These features allow you to edit (change) the on-screen text without having to retype the whole document.

It is easy to get muddled over the meaning of the terms *cut*, *copy*, *move* and *paste*.

- **Cut** ✄ deletes the selected text and places a copy onto the *Clipboard* (a special part of the computer's memory) from where it can be retrieved.

- **Copy** 📋 copies the selected text onto the Clipboard, without deleting it. It can then be placed elsewhere in the document.

- **Move** means to cut selected text from one place and immediately paste that text elsewhere in the document.

- **Paste** 📋 inserts the text from the Clipboard elsewhere in the document, or even into a different document.

Formatting text

Font type and font size

A *font* is the name given to describe the style of typeface you are using. Two popular styles frequently used in the preparation of business documents are Times New Roman and Arial.

- This is an example of Times New Roman. It is referred to as a *serif* font because of the 'little feet' at the bottom of each letter.
- This is an example of Arial. It is referred to as a *sans serif* font because it does not have the 'little feet'.

Your default font will generally be Times New Roman, and the size is likely to be somewhere between 10 and 12 points. *Point* refers to the size of the character – the higher the number, the larger the font. A point size of 72 would give you a letter approximately 2.54 cm (1 inch) tall.

It is important that you choose a style and size of font to suit the task you are doing. Some font styles are rather elaborate and difficult to read, so you should use them with care. Similarly, choose a suitable font size.

For example, text contained in a poster will probably need to be quite large so that it is eye-catching and can be read from a distance. In contrast, a leaflet or newsletter might contain fairly large headings, with the body text printed in a smaller font.

TiP

Most business documents will rely on the use of capital letters, bold, underline or italic to emphasise any key points to be made.

Bold, underline and italic

In addition to using different font styles and sizes, you can also use **bold**, underline or *italic* to emphasise text. ***However, it is not generally a good idea to apply all three to the same text!***

Using colour

Colour is not generally applied to text in standard documents, such as letters or reports, but it can be very effective in leaflets, posters, flyers, etc. However, for it to be effective you should use it in moderation – otherwise your document will be messy, difficult to read and therefore may not be fit for the purpose intended.

Your word-processing software has a selection of standard colours (Figure 19(a)). It also has the facility to customise a colour

(Figure 19(b)), which can be useful if you need to match an existing colour. For example, some companies have 'corporate colours' that they replicate on their logo, advertising material, brochures, on-screen presentations, and so on.

(a)
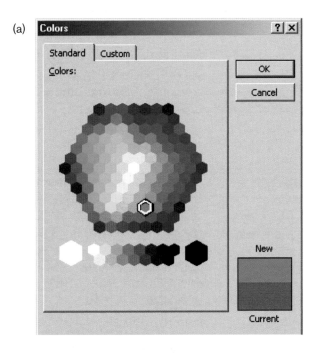

(b)
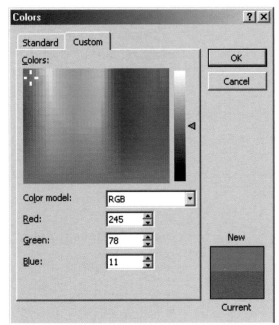

Figure 19 (a) The standard colour palette and (b) the custom colour option

Using paragraph formatting features

Alignment

Usually your text automatically lines up against the left-hand margin, leaving the right-hand margin uneven (or *ragged*). This is because the default style of paragraph alignment is *left aligned*. Word-processing software offers you alternative styles of paragraph format, such as *justified* or *centred*.

Here the text is **left-aligned** and the right-hand margin is uneven or *ragged*. This is the default setting. The spaces between words are equal.	Here the alignment is **justified** and both the left and right margins are straight. The program automatically adjusts the spaces between words to distribute the text evenly between the margins. The spaces between words are not equal.	Here the text is **centred** between the margins. This is generally used for presentation and display rather than letters or notes.

Figure 20 Text left-aligned, justified and centred

The default setting of left alignment is quite acceptable for most text-based documents and on-screen presentations. However, there are occasions when justified alignment may be more appropriate. For example, a newsletter produced in columns looks more professional with justified margins (Figure 21(a)). Lines of text in a poster may be centred between the margins; in order to balance the presentation, paragraphs in the same poster look neater if they are justified, leaving an equal space against the left and the right margins (Figure 21(b)).

Bullets and numbering

Bullet points are used to make items in a list stand out. The standard bullet point is represented by the character symbol ●, but you can choose different symbols. For example

- a bullet point chosen from the standard selection in Bullets and Numbering (Figure 23)
- a bullet point customised from a selection in Wingdings
- a bullet point customised from a selection in Webdings.

(a)

(b)

Figure 21 (a) A newsletter and (b) a poster

As you can see from these examples, the standard bullet point is the clearest and is the style most commonly used in the preparation of paper-based documents.

Sometimes different styles of bullet point are used to indicate different 'levels' in a list. The list that follows is an example.

- Fully-inclusive holidays
  » Full-board
  » Half-board
  » Self-catering
- Activity holidays
  » Skiing
  » Water sports

The important thing to remember is that you should use a style of bullet that is fit for the purpose of the document. Use the decorative or picture bullets only for more informal documents, on-screen presentations or web pages.

Sometimes it is preferable to number a list rather than using bullets. The *Numbered* tab, which you can see at the top of Figure 23, offers a variety of options – such as **1, 2, 3** or **a), b), c)**. Figure 22 shows both bullets and numbering used in a document.

TRAVELBUG CRUISE CLUB

Our Cruise Club is totally independent and we can therefore offer a varied range of cruises from all leading cruise line operators. Our knowledgeable staff are on hand 7 days a week to help you plan the perfect trip.

Membership is free and Cruise Club members have the following advantages:

- 2 for 1 cruise offers
- Services of our specially trained staff
- Monthly newsletter with the latest information and offers
- Cabin upgrades and on-ship spending vouchers
- Exclusive visits to see the ships whilst docked in the UK
- Specially discounted rates

How to join

1. Call us for an application form or download from our web site
2. Complete the application form
3. Return to us in Whychton
4. Your application membership pack will be sent out by return post.

Figure 22 The use of bullet points and numbering

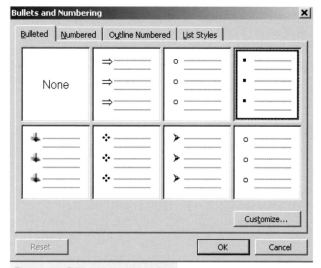

Figure 23 Different bullet styles

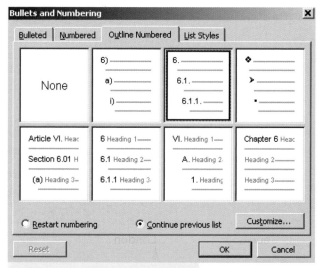

Figure 24 A section numbering style

Tabs

The **Tab** key is located to the left of the letter Q on your keyboard. Every time you press the **Tab** key the cursor jumps across the page. The **Tab** key is used to place and align text on the page.

In the example in Figure 25, tabs have been used to position the three columns headed 'Coffee', 'Beer' and 'Meal for 2 people'.

Approximate costs					
→	→	*Coffee*	→	*Beer*→ →	*Meal for 2 people*
France→	→	£1.40→	→	£2.50→ →	£44.00
Italy→	→	£0.70→	→	£1.45→ →	£27.50
Spain→	→	£0.80→	→	£1.00→ →	£20.00

Figure 25 The use of tabs to form a simple table

Figure 26 illustrates Microsoft Word's ruler. If you look closely you will see very feint vertical marks along the bottom edge.

These marks represent the default tabulation settings (or tab settings) and are positioned at 1.27 cm (½-inch) intervals.

Figure 26 The ruler and default tab settings

Tab styles

The default tab settings are known as *left tabs* because the text is left-aligned with the tab. There are three other useful tab settings illustrated in Figure 27:

- right – text is right-aligned with the tab
- centre – text extends either side of the tab setting
- decimal – text before the decimal point extends left and after the decimal point extends right.

Figure 27 Column alignments in a tabbed table

4 STAR OFFERS				
Date	*Place*	*Hotel*	*No. of Nights*	*Cost*
10 October	Derby	Broadmeadow	4	£210
16 November	Chester	Lodge Gate	6	£249.50
23 November	Edinburgh	Monterry	5	£265.99
6 December	Brighton	Hurlingham Park	1	£75
13 December	London	Stretford	2	£150
↑		↑	↑	↑
Left		Right	Centred	Decimal

Indents

Paragraph indents can be used to make paragraphs stand out. Indents can be applied by selecting **Format, Paragraph** from the menu, or by dragging the indent markers on the horizontal ruler (Figure 28). Figure 29 shows some examples of paragraph indents in a newsletter.

← First line indent

← Hanging indent

Figure 28 Indent markers on the horizontal ruler

TRAVELBUG CITY BREAKS

Luxury coach tours
5 nights from £139.00
Nationwide pick-up points

Take advantage of the special deals we have negotiated with Whychton Coach Company and travel to one of your favourite European destinations by luxury, air-conditioned coach. All coaches have facilities for refreshments, WCs and videos and for your safety 2 qualified coach drivers accompany all journeys.

Paris is a shopper's paradise and is renowned as a city of culture. Your first stop must be the Louvre to see the Mona Lisa and Pei's glass pyramid. If Art Nouveau and Impressionism are more to your liking, then a visit to the Musée d'Orsay is a must.

Barcelona is the place to go to see the unmistakable works of Gaudi. Visit his distinctive houses in the Paseo de Gracia or the awe-inspiring church of the Sagrada Familia. Stroll through Las Ramblas to the busy port and Gothic Quarter.

Venice is a beautiful city and an unforgettable experience. Everybody goes about their business on the broad canals and narrow waterways. The famous Rialto Bridge is lined with shops selling glass, lace and carnival masks. St. Mark's Square is the place to be seen.

Florence represents the cultural heart of Italy and is packed with splendid architecture and fine art, including Michelangelo's famous statue of David and paintings by Raphael, Botticelli and da Vinci.

17 London Road   Whychton   TO9 3WN

Tel: 543 2134 5678   www.travelbug.co.uk
Fax: 543 2134 5679   Email: info@travelbug.co.uk

Figure 29 Examples of indents on a newsletter

Line spacing

The distance between each line of text in a paragraph is known as the *line spacing*. The default setting in your word processor will produce text in single line spacing. However, there are occasions when it is useful to leave a larger space between the lines. For instance, you might want to make a section of text stand out, or to leave room between lines to make handwritten notes (Figure 30).

This example of text has been produced in **single line spacing**. This is the default setting and is probably the most commonly used.	This example shows text in **one and a half line spacing**. It is used to make sections of text stand out and therefore become easier to read.	This example shows text produced in **double line spacing**. It is particularly effective if you wish to write notes between the lines, as on a draft document.

Figure 30 Examples of single, 1.5 and double line spacings

Tables

We have already seen on page 93 that columns of data can be presented tidily by using tabs. A more powerful option, which can help you to display information effectively, is to use *tables*.

Tables consist of rows and columns that form individual boxes (or *cells*) – rather like a spreadsheet. Each individual cell may contain any amount of text, a picture or even a mathematical formula.

Figure 31 shows a basic table prepared by Travelbug to tell their customers the price of overnight accommodation in France.

Hotel stopovers
Prices shown in £s per room

Category	Accommodation	Weekend	Extra Child
Room only	Twin/double	£35	N/A
	Family	£45	2 free
Bed and Breakfast	Twin/double	£55	1 free
	Family	£75	2 free

Figure 31 A basic table layout

Borders and shading in tables

There are additional features in your word-processing software that give you the opportunity to enhance or improve the general appearance of a table. For example you can apply borders and shading, and introduce colour.

Figure 32 shows the table from Figure 31 with the addition of some features.

Hotel stopovers
Prices shown in £s per room

Category	Accommodation	Weekend	Extra Child
Room only	Twin/double	£35	N/A
	Family	£45	2 free
Bed and Breakfast	Twin/double	£55	1 free
	Family	£75	2 free

Figure 32 The table in Figure 31 with the addition of an outside border, shading and colour

Modifying row and column sizes

The columns showing prices for weekends and extra child details are wider than they need to be. The column width can be modified by picking up the arrow shape +‖+ and dragging the column border to a new position. The effect of doing this is shown in Figure 33.

You can also modify the row height by picking up the arrow shape ÷ and dragging up or down.

Hotel stopovers
Prices shown in £s per room

Category	Accommodation	Weekend	Extra Child
Room only	Twin/double	£35	N/A
	Family	£45	2 free
Bed and Breakfast	Twin/double	£55	1 free
	Family	£75	2 free

Figure 33 Compare this with Figure 43: the column widths have been adjusted

Inserting and deleting columns and rows

In Figure 34 an extra column has been added to the table to show the midweek prices. Extra rows have also been added to enable the headings to become part of the table and to separate the two different categories of accommodation. Extra columns and rows can be inserted by selecting **Table, Insert**.

HOTEL STOPOVERS				
ALL OFFERS SUBJECT TO AVAILABILITY				
Category	**Accommodation**	**Weekend**	**Midweek**	**Extra Child**
Room only	Twin/double	£35	£35	N/A
	Family	£45	£45	2 free
Bed and Breakfast	Twin/double	£55	£60	1 free
	Family	£75	£80	2 free
Prices shown in £s per room				

Figure 34 Compare this with Figure 33: the table now has an extra column, and four additional rows have been inserted

Creating, selecting and inserting components

Types of components

Many of the publications you produce may comprise more than one component, such as

- images
- lines and simple shapes
- tick boxes
- comments
- hyperlinks.

You must remember that whatever you include in a publication should be there because it improves the effectiveness of that publication and not just because you felt like putting something in.

Images

Images used to illustrate a topic can certainly help people to understand the topic and generally make a publication more interesting for the reader. However, you should only include images that are fully relevant to the topic and that serve a purpose by being included.

You may search through a picture library to find a relevant image, but you must be very careful to obtain permission to use the image if it is protected by copyright laws. Images taken with a digital camera can be inserted directly from the camera. Photographs and images in books and magazines or photographs you take with a traditional camera will need to be scanned into the document.

Remember to record full details of the sources of all images you include in your work.

Lines and simple shapes

You will find it very useful to investigate the *Drawing* toolbar and to experiment with the variety of lines and shapes that are available for you to use. Move your mouse pointer over the toolbar to see the names given to the various tools.

Figure 35 The Drawing toolbar

The arrow tool is very helpful if you need to label a diagram. Other shapes can be built up to form a variety of objects.

The image of the train in Figure 36 is made up from approximately 30 lines and shapes. Some have been copied, some have been filled with colour and all lines have been made thicker.

Figure 36 A train constructed from simple shapes

> **✓ TiP**
>
> If you find it difficult to place one object in front of another, use the **Order** option in the Draw menu to send an object in front or behind another.

> **✓ TiP**
>
> If you hold down the **Ctrl** key whilst using the arrow directional keys, you can 'nudge' the lines and shapes across the screen in very small steps. If you increase the zoom on the screen to 200% or 500% you can see very clearly when lines and shapes are in the right position.

Grouping and ungrouping

You have produced an image that is made up of many different lines and shapes. If you try to move the image, it will be impossible to keep all the shapes together unless you *group* them. Grouping allows you to treat a group of objects as a single object that can be rotated, resized or flipped. Any group of images can be *ungrouped* and treated as a number of separate objects again.

Borders

You already know how to place a border round text (see page 96) but you may sometimes wish to place a border round a full page, for example on a poster. You will find a wide choice of artwork and line styles available in **Format, Borders and Shading, Page Border**.

You can use the *Rectangle* tool to place a border around an image or object. Draw the rectangle shape over the image with the **Rectangle** tool on the *Drawing* toolbar. The image will disappear under the rectangle but you can double-click the rectangle, remove the 'Fill' colour and choose a suitable line style and colour. You can resize the rectangle as necessary.

Using images and objects

Many documents and on-screen presentations contain illustrations. These are included to help simplify the information contained in the text, or simply to provide decoration. In this book many types of illustrations have been used: pictures, photographs, charts, graphs, screen prints, text boxes, clip art, WordArt, etc. Collectively these are referred to as *objects*.

Sensible use of images will greatly improve a document and make it more appealing to the target audience. Figure 37 is an example.

An image or object can be inserted by selecting the **Insert** drop-down menu.

TRAVELBUG WATER SPORTS

Departing 20 June
Travel by luxury coach from Whychton
14 nights half-board accommodation in 3-star hotel

A fantastic opportunity to experience a variety of water sports including:

Windsurfing

Learn the basics of windsurfing, including rigging, equipment care, sailing across the wind and self-rescue. Advanced skills include instruction on topics such as water starts, harness use, and shortboards. Wetsuits and windsurfing gear are provided.

Snorkelling

For many people the water's edge is the limit of their activity and knowledge. Learn to snorkel and enjoy the world that lurks beneath the surface of the water. You will be introduced to snorkelling equipment including the mask, flippers and snorkel and will learn how to move safely in the water.

Water skiing

If you are a complete beginner you will start on land and learn the fundamentals of body position, technique and safety. Once you and your instructor are confident in understanding the basics on land, you move to the water. First-time skiers will use the static boom on the side of the boat to offer stability. When you have mastered the boom, then it is off to the long line! Who knows, you may be able to Slalom, Trick or Jump before the end of your holiday!

£585 per person
(based on two people sharing a room and fully inclusive of all equipment and tuition)

17 London Road   Whychton   TO9 3WN

Tel: 543 2134 5678   www.travelbug.co.uk
Fax: 543 2134 5679   Email: info@travelbug.co.uk

Figure 37 Sensible use of images will greatly improve a document and make it more appealing to the target audience

Positioning images and objects

Any image in a document should be there to serve a specific purpose, and its positioning is therefore very important. The position of the cursor when the image is inserted into the document will determine where it appears. Figure 38 shows an example. After the image was inserted, the text was pushed down the page.

Windsurfing
Learn the basics of windsurfing, including rigging, equipment care, sailing across the wind and self-rescue. Advanced skills include instruction on topics such as water starts, harness use, and shortboards. Wetsuits and windsurfing gear are provided.

Windsurfing

Learn the basics of windsurfing, including rigging, equipment care, sailing across the wind and self-rescue. Advanced skills include instruction on topics such as water starts, harness use, and shortboards. Wetsuits and windsurfing gear are provided.

Figure 38 An example of text before the insertion of an image, and the same text with an image: notice how the text has been pushed down the page

Wrapping text around an image

As you can see from Figure 38, although an image has been placed in the document, its inclusion does not really improve the document because it is positioned on the left with empty space at the side. In order to make the document more interesting and the image more manageable, you can format the image and select a *wrapping style* to improve the layout. Figure 39 shows two examples.

Windsurfing
Learn the basics of windsurfing, including rigging, equipment care, sailing across the wind and self-rescue. Advanced skills include instruction on topics such as water starts, harness use, and shortboards. Wetsuits and windsurfing gear are provided.

Windsurfing
Learn the basics of windsurfing, including rigging, equipment care, sailing across the wind and self-rescue. Advanced skills include instruction on topics such as water starts, harness use, and shortboards. Wetsuits and windsurfing gear are provided.

Figure 39 A 'square' wrapping style, and a 'tight' wrapping style in which the text hugs the edges of the image

TiP

You should make sure that any images you include in your documents improve their appearance and are not included just because you fancied putting them in!

Wrapping of an image can be achieved by selecting the **Format, Picture** (or **Object**), **Layout** drop-down menu.

Images and objects that are formatted to wrap text can also be moved more easily within the document by clicking and dragging them to a new position.

TiP

Take care when resizing an image. If you select a corner handle the image will be resized while retaining its true proportions. However, if you select a centre top, bottom or side handle you could end up with images looking like Figure 40.

Sizing an image

Sometimes the size of an image or object should be restricted because of the amount of available space on the page. Any object can be resized by selecting the object to show the selection handles and dragging one of the handles with the mouse.

Figure 40 Images that have been resized without retaining the original proportions: inclusion of these images would not improve a document!

Cropping an image

From time to time an image we require might be part of a larger image. We can 'crop' the unwanted area from the image by using the **Crop** tool in the *Picture* toolbar.

(a)

(c)

(b)

Figure 41 (a) A clip art image of a sea shell on the beach. (b) The sea shell has been 'cropped', removing the surrounding area. (c) The cropped image can then be resized if necessary.

Using spelling and grammar checkers

The importance of using spelling and grammar checkers has been stressed in Standard Ways of Working (on page 75).

As a student on an IT course, you have no excuse for submitting work which contains spelling and grammar errors – the tools to help you check the accuracy of your work are at your fingertips. Make sure nothing is left unchecked!

Presentation software

Presentations can be in a variety of forms. At school or college your teacher may present information using an overhead projector and slides. A cinema's booking office could use a computer to show a series of pictures about forthcoming films; this would also be a presentation.

Travelbug could use Microsoft PowerPoint to create a presentation of its products and services to show prospective clients. You will probably be asked to deliver a PowerPoint presentation during your school or college studies, and you may even be asked to deliver one when you go for a job interview. Learning to use PowerPoint effectively will therefore provide you with a skill that you will find useful both now and in the future.

LEARNING OUTCOMES

You need to learn about

✓ designing and creating the structure and navigating route of a presentation

✓ selecting and creating colour schemes

✓ creating and selecting components – text and graphics

✓ using master slides and templates

✓ using frames

✓ editing text – fonts, aligning, bullets and numbering, line spacing

✓ editing graphics – aligning, rotating, flipping, cropping and resizing, changing colour and resolution

✓ **using transparency**

✓ **optimising file size**

✓ adding lines and simple shapes

✓ using text wrap

✓ creating slide transitions

✓ producing speaker notes and handouts to accompany slide shows

✓ **rehearsing and checking timings of a slideshow**

✓ **packing a presentation for transfer to another computer.**

Designing and creating the structure of a presentation

A PowerPoint presentation is made up of individual parts, called *slides*. You can build up a series of slides to create a presentation. Often you will see a PowerPoint presentation displayed through a digital projector on to a screen. You may have experienced your teachers doing this in lessons. Alternatively the presentation could be shown on a computer screen as an on-screen display – for example the presentation about forthcoming films.

It is essential to plan a presentation before creating it. This includes creating a *storyboard* that shows the layout and content of each individual slide. You should also plan the structure of the presentation, making sure you show the *navigation route*.

Using wizards, templates and master slides

Wizards

You may already be familiar with the wizards used in applications such as Microsoft Access. PowerPoint also provides a wizard to simplify the creation of a presentation. In PowerPoint this is called the *AutoContent Wizard* (Figure 42). If you want to create a presentation quickly and be guided step by step, this is an ideal feature.

Figure 42 The AutoContent Wizard

Templates

PowerPoint also makes creating a presentation easier by providing a wide range of design and presentation *templates* (Figure 43).

A design template is a file that has been designed with special backgrounds and layouts ready to use. It includes styles for the type and size of bullets and fonts. Using a design template is another timesaving feature, and, if you are not particularly artistic, can be invaluable.

The design template always has two slide designs – one used for the title slide (the first slide in a presentation) and one used for the remaining slides. This means that the title slide will have a slightly different design from the remaining slides in a presentation.

Presentation templates are pre-structured presentations that you can choose to suit a specific purpose (Figure 44). For example, Thomas Tripp from Travelbug could use the marketing plan template to create a presentation to market holidays to prospective clients.

If you don't want to use one of the templates supplied with PowerPoint, you can find different themes and design templates on the Internet, many of which you can download without charge.

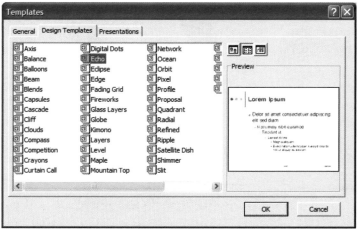

Figure 43 The Design Templates window

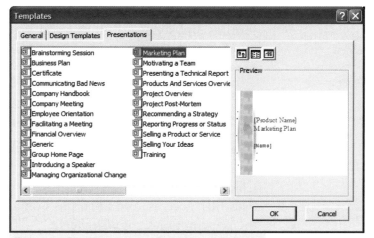

Figure 44 The Presentation Templates window

Jargon buster

Default settings are the standard settings you find each time you use a software package. For example, the default orientation setting for a slide in PowerPoint is landscape.

Jargon buster

An **accent** is a block of colour used for highlighting.

The slide master

Each design template comes with a *slide master* on which you can put any graphics or text that you want to appear on every slide, and an optional *title master* where you can make changes to slides in your presentation that use the title slide layout. For example, Travelbug can put their logo on the slide master so that it will appear on every slide (Figure 45).

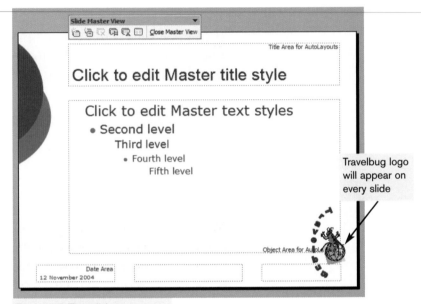

Figure 45 The slide master

If you wish to divide your presentation into several sections, each starting with a title slide, you can create a separate title master. You edit the slide masters and title masters to set the default text formats and styles for all slides in your presentation. You can also number each slide, include a footer, and show the date a presentation was created.

Selecting and creating colour schemes

Colours are used for the background, text and lines, shadows, title text, fills, accents, and hyperlinks on a slide. Together they are called the presentation's *colour scheme*.

If you choose to use an existing template for your presentation, one useful feature is that if you do not like the colour scheme then you can change it easily (Figure 46). A design template will include a default colour scheme for the presentation, together with additional alternative schemes to choose from.

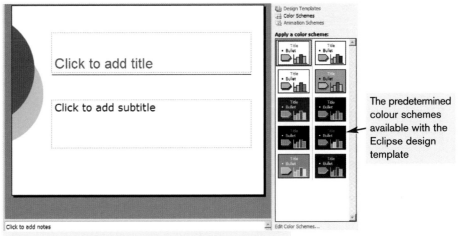

Figure 46 Changing the colours of a design template

You can even change certain aspects of a default colour scheme. For example, suppose Thomas Tripp of Travelbug wants to create a PowerPoint presentation to be shown at a travel exhibition. Because he wishes to ensure that the colour scheme uses the Travelbug house style, he can choose to modify the design template to reflect this. As PowerPoint allows you to change the colour for all or only certain elements of a colour scheme, Thomas Tripp could change the colour for text and lines, but leave the other elements unchanged (Figure 47).

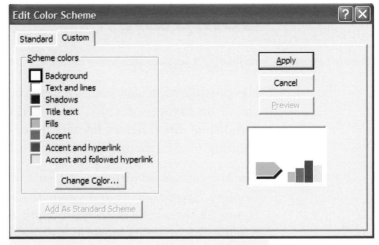

Figure 47 Dialogue box to edit a colour scheme

Colour schemes can be applied to one slide, selected slides or the entire presentation. If you are creating the template yourself for a presentation, you can apply colour schemes in the same way you would if you were using an existing design template.

Viewing your slides

There are three main ways of viewing slides.

- *Normal view*. This is the main view used for editing. It displays three areas. On the left-hand side there are tabs which alternate between slide and outline view, the slide pane and the notes pane. You can adjust the pane sizes by dragging the pane borders. Slides are displayed individually and you can work on the slides in this view. The notes pane allows you to enter notes that you want to make about a slide, which will assist you when making a presentation.

- *Slide sorter*. This allows you to view all the slides in miniature form. Not only can you delete slides, change the order of slides or insert new slides in slide sorter view, you can also copy existing slides and paste them into the desired positions very easily.

- *Slide show*. You can view your presentation by clicking on the Slide Show icon (⌨) in the bottom-left.

Creating, selecting and using text and graphics components

Slide layout

It is possible to design your own slides from scratch, or to choose one of the layouts provided by PowerPoint. The *Slide Layout* option will display the different layouts from which you can choose. By pointing to each picture you can see a description of the slide.

The layouts are divided into different categories – *Text Layouts*, *Content Layouts*, *Text and Content Layouts*, and *Other Layouts*. Figure 48 shows the different layouts, which allow a range of components to be included in a slide: text (including bulleted lists), charts, graphics, sound and video (referred to as *media clips*).

PowerPoint allows you to click on icons to add the appropriate content for a slide. For example, Thomas Tripp would click on the **Insert Picture** icon to insert a digital photograph in his presentation (Figure 49).

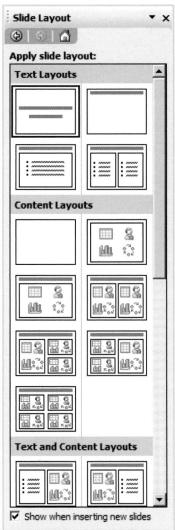

Figure 48 The slide layout options

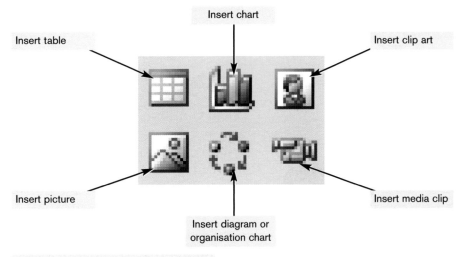

Figure 49 The components available

Using boxes/frames

The design of each slide is broken down into areas called *boxes* (sometimes referred to as *frames*). Each box/frame holds an object. The object can be, for example, a list, a title, text, a piece of clip art, or a chart. A box can be copied, moved and resized in the same way as you would a piece of clip art, and can even be rotated.

You can either use one of the slide layouts provided by PowerPoint, or start with a blank slide and insert objects wherever you like on the slide. These objects are automatically held in boxes.

Editing text on slides

Many of the features that you may be familiar with from using other applications – such as Word or CorelDraw – can be applied to text in PowerPoint by selecting **Format** on the *Standard* toolbar (Figure 50). You can format text to bold, italic, underlined or shadowed, and align text to the left, centre, right or justify it. You can change the colour of text and choose from a wide range of fonts and sizes.

Skills check

Refer to page 88 for information on formatting text.

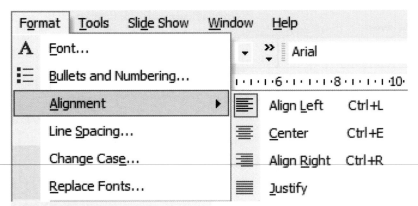

Figure 50 The Format menu options

Using the *Line Spacing* dialogue box from the **Format** menu, you can adjust the line spacing, not only between lines, but before and after paragraphs, just as you can in Word.

Some slide layouts allow for text areas that are specifically for lists. When you start typing in one of these, a bullet will appear before the text. You can format the bullets to different styles, in the same way as you would in Word. You can also create numbered lists, again exactly as you would in Word.

Editing and using graphics

The graphics you include in a presentation could be a photograph taken with a digital camera, some clip art, a drawing, or perhaps a logo such as the Travelbug logo.

 Think it over ...

Have you heard the expression 'a picture is worth a thousand words'? Including graphics in a presentation will ensure that it is more interesting, often easier to understand and – most importantly – more memorable.

You may decide to use a specialist editing package or the *Picture* toolbar in PowerPoint (Figure 51) to edit a graphic. For instance, Thomas Tripp might decide to adjust the colour and resolution of a digital photograph using Coral Photo-Paint or Microsoft Photo Editor rather than use the more limited options supplied with PowerPoint.

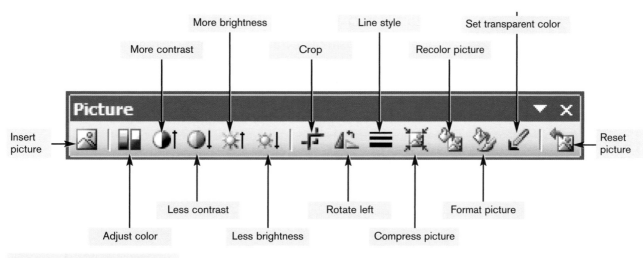

Figure 51 The Picture toolbar

Transparency

You can use the **Set transparent color** tool to remove one colour from your picture and let whatever is behind it show through. This is typically used to remove a coloured background from an image. Some graphics formats, such as GIF, can be saved with transparent areas.

Inserting clip art

Inserting clip art into a slide is very easy – PowerPoint comes with a gallery (Figure 52). If you have an Internet connection, you can click on the **Clips Online** button on the menu bar. This will connect you to Microsoft's database of free clip art. You can also buy a CD-ROM full of pictures you can use.

Skills check

Positioning, cropping, resizing, grouping and borders are controlled in the same way as in Microsoft Word. See page 97 for details.

TiP

*The Internet is a wonderful source of graphics that you can download. However, whatever the source of your graphics, it is important to remember to check that you are allowed to use them. If the creator says that a picture is in the **public domain**, then this means that the copyright on it has been waived and it is free to use. Otherwise you may have to get permission.*

Click here to access Microsoft's Clips Online.

Figure 52 The Microsoft Clip Organizer

Optimising the file size of an image

One aspect you should be aware of when creating a PowerPoint presentation that includes a range of images is that the overall file size can quickly become huge. A very useful feature of PowerPoint 2002 and later versions is that you can compress images and remove unneeded data (Figure 53). For example, it will delete cropped areas of pictures from the file.

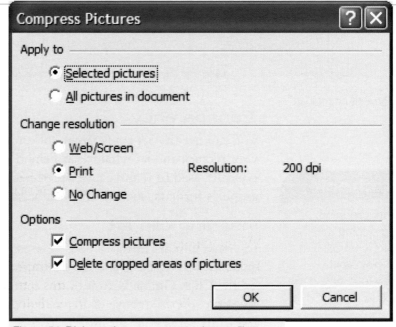

Figure 53 Dialogue box to compress image files

Adding more features

Inserting lines and simple shapes

There are many ways to improve a PowerPoint presentation. You can add lines and shapes, fill the shapes with colour, outline them and make them look three-dimensional. The **Autoshapes** button on the *Drawing* toolbar provides a menu of types of shape, each type having its own sub-menu showing all the shapes available. There are over 150 different shapes to choose from!

Callouts are designed to hold text within the shape. They can be simple boxes with lines pointing from them or 'word and thought' balloons (Figure 54). As with other text boxes, you can resize a callout, and rotate and format its text.

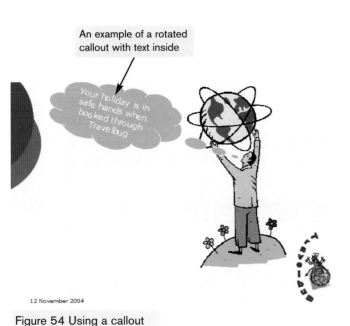

An example of a rotated callout with text inside

Your holiday is in safe hands when booked through Travelbug

12 November 2004

Figure 54 Using a callout

Click here to turn off word wrap

Format AutoShape

Colors and Lines | Size | Position | Picture | Text Box | Web

Text anchor point: Middle

Internal margin

Left: 0.25 cm   Top: 0.13 cm

Right: 0.25 cm   Bottom: 0.13 cm

☑ Word wrap text in AutoShape
☐ Resize AutoShape to fit text
☐ Rotate text within AutoShape by 90°

OK | Cancel | Preview

Figure 55 Turning off text wrapping in an AutoShape

Wrapping text

By default, text inside a shape is wrapped so that it does not spill over the border. If you need to turn this setting off, double-click the shape to display the *Format AutoShape* dialogue, and untick the **Word wrap text in AutoShape** option in the *Text Box* tab (see Figure 55).

Inserting WordArt

As with other Microsoft Office programs, such as Word and Excel, you can use the *WordArt* tool to create a logo, or to make text more interesting. One very useful feature is that a design you make in one Office application can be used in any of them. Therefore, if you have used WordArt in Word to create a logo, for example, you could use the same logo in a PowerPoint presentation.

? Think it over ...

Look at the two sample slides in Figure 56. One of these has WordArt for the text, and the other does not. Which do you think looks more appealing?

WordArt used for text

Book Your Holiday With Travelbug Now

Book Your Holiday With Travelbug Now!

Normal text

Figure 56 A comparison of WordArt and normal text

Navigation between slides

The default navigation route of a PowerPoint presentation is linear. However, the *Action Settings* feature will allow you to link to another slide further on in your presentation, to another PowerPoint presentation, to a file, or even to a website (Figure 57). You can add settings to text or to an object in your presentation.

The **AutoShapes** button on the *Drawing* toolbar also includes a number of action buttons which you can include on slides. This feature can be useful when you wish to move to another part of your presentation, or even to another file.

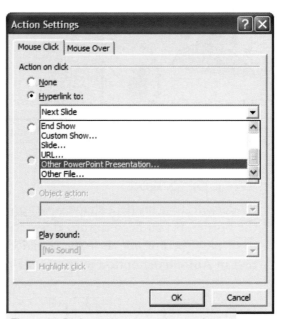

Figure 57 Dialogue box for action settings

Thomas Tripp has built a slide which gives examples of what Travelbug can offer their customers (Figure 58). He has also used the action settings facility to link to slides in a different part of the presentation (Figure 59).

Finally, Thomas has included an action button which he has set to return to the last slide viewed.

By clicking here, the presentation moves out of sequence to show the Cruise Club slide.

Figure 58 Slide with links to different parts of the presentation

Figure 59 Adding a hyperlink from slide 4 to slide 9

Creating slide transitions

When you are showing a PowerPoint presentation and move from one slide to the next, this is referred to as *slide transition*. When you design a PowerPoint presentation, each slide has a transition associated with it. The transition will tell PowerPoint how to change the display from one slide to the next.

The latest version of PowerPoint offers over 50 different transition styles from which to choose – examples are *cut*, *dissolve* and *wipe right*. If you prefer, you can choose a random transition, so that a different style and direction will be used each time you move on to a new slide.

You can adjust the speed for slide transitions to be slow, medium or fast. Furthermore, you can make the transitions occur on the click of a mouse or automatically after a set period of time. You can even set up a presentation to show continuously until the **Esc** key is pressed. Thomas Tripp could use this feature for the slide show he wants to show at a travel exhibition.

Rehearsing and checking timings of a slideshow

Before you give a PowerPoint presentation, it is important to rehearse and check the timings carefully. Consider these points:

- Is the timing correct between objects appearing on a slide?
- Are the transition timings for slides correct?
- If you have set timings for objects or slide transitions to appear automatically, rather than on the click of a mouse, have you allowed sufficient time to say everything you want to at a reasonable pace?
- Will the audience have time to read text before the next bullet point or object appears?

Producing speaker notes and handouts

Speaker notes

PowerPoint has a facility called *speaker notes*, which allows you to include notes with a slide. This feature gives a text display for each slide. You can add anything you want in these notes (Figure 60). For example, Thomas Tripp of Travelbug can use this facility to give extra details on the different cities shown on the City Break slide of his presentation. The slide, together with the notes, can be printed out to give to prospective customers.

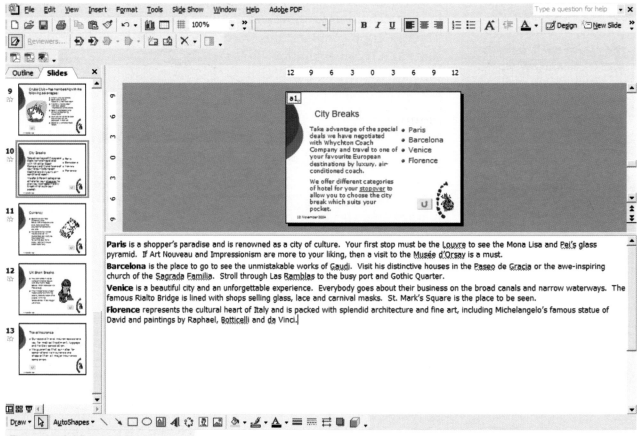

Figure 60 Adding speaker notes

Handouts

Another feature of PowerPoint is that you can print a presentation as a *handout*. You have the option of choosing how many slides per page to print, up to a maximum of nine. If you choose three slides per page, as shown in Figure 61, PowerPoint automatically adds lines to the right-hand side of each slide so that the audience can write their own notes during the presentation.

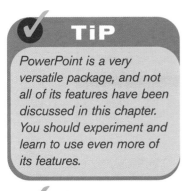

TiP

PowerPoint is a very versatile package, and not all of its features have been discussed in this chapter. You should experiment and learn to use even more of its features.

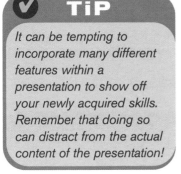

TiP

It can be tempting to incorporate many different features within a presentation to show off your newly acquired skills. Remember that doing so can distract from the actual content of the presentation!

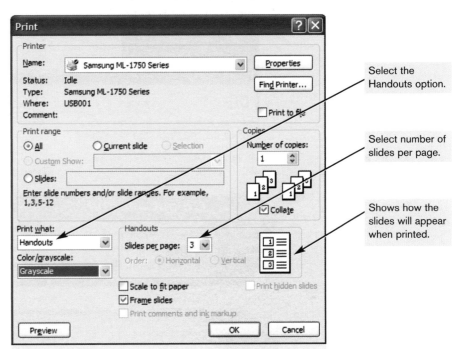

Select the Handouts option.

Select number of slides per page.

Shows how the slides will appear when printed.

Figure 61 Options to choose from when printing handouts

Pack and Go

The travel exhibition where Travelbug have booked a stand is providing computers for exibitors. However, Thomas Tripp is uncertain whether PowerPoint will be installed on the machines.

Rather than take the risk that PowerPoint will not be available, Thomas decides to use the *Pack and Go* feature in PowerPoint. By choosing the option to include the PowerPoint Viewer (Figure 62), Thomas will be able to run the presentation in PowerPoint format even if the computer does not have the program installed.

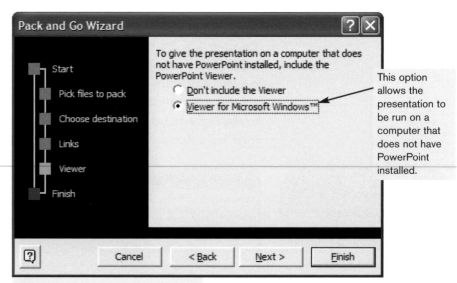

Figure 62 The Pack and Go Wizard

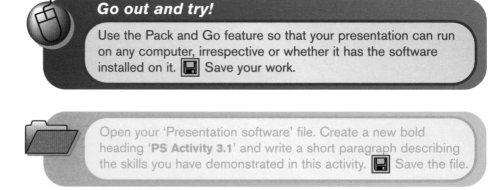

Go out and try!

Use the Pack and Go feature so that your presentation can run on any computer, irrespective or whether it has the software installed on it. 💾 Save your work.

Open your 'Presentation software' file. Create a new bold heading '**PS Activity 3.1**' and write a short paragraph describing the skills you have demonstrated in this activity. 💾 Save the file.

Artwork and imaging software

With the power of the modern computer and its ability to select, capture or modify images, you don't have to be an artist to create interesting, imaginative and effective artwork. This book contains a variety of images and artwork: some are screen prints of the computer, some are photographs and some are drawings; some are in black and white and others are in colour. They are all intended to make the book more interesting to read and to help you understand the topics you are studying.

Sometimes a graphical representation can stand alone without any words at all, and at other times the image makes the words much easier to understand, or vice versa. For example, it would be very difficult to describe charts and graphs just in words.

LEARNING OUTCOMES

You need to learn about

✓ selecting and capturing images

✓ modifying images

✓ preparing images for screen and print

✓ choosing appropriate resolutions and file formats.

Selecting and capturing images

Skills check ▷▷

See page 141 for information about downloading images from the Internet.

There are several ways to select and capture images that can then be used to improve the presentation of your e-portfolios. You can select images that are already prepared (such as clip art), create your own using drawing tools, or acquire images electronically by scanning them or downloading photos from a digital camera.

Using clip art and library images

Images can be imported into your documents from libraries on CD–ROM, the Internet, or packaged with the software. Word contains its own selection of images, but far more are available from other sources. Many of these clip art libraries are free.

You can enter a topic in the search field and the computer will find relevant pictures available in the library. A search for 'Australia' produced many results. The search was then refined to show only images first in clip art format and then in photo format. Two images of Sydney Opera House were selected, and copied and pasted into the text for this chapter (Figure 63).

Figure 63 Sydney Opera House in clip art format, and in photo format

Figure 64 A scanner

Using a scanner

A *scanner* allows you to add pictures from other sources – such as a magazine, book or photograph – into your documents (Figure 64). The scanner reads the information and converts it into digital format.

The document is placed inside the scanner. Once the image has been scanned, it can be stored in the computer and used intact or edited as required. The scanned image is stored in picture format, even if it is text – unless text recognition software is available, in which case the text is stored as data that can be edited using a word processor.

Using a digital microscope

A digital microscope includes a built-in camera that makes it possible to view specimens on a TV or a computer. Magnified images can be saved on to the computer, together with live video or time-lapsed film of objects such as insects or plant life. It is then possible to manipulate these images by adding text or special effects, or to incorporate them into another document.

Go out and try!

If you have access to a digital microscope (Figure 65), select an item, such as a leaf or flower, to study under the microscope. Transfer the image to the computer. Write a short description of the detail, not visible to the naked eye, that you can observe through the microscope. Add the image to your description and label the different parts of the image.

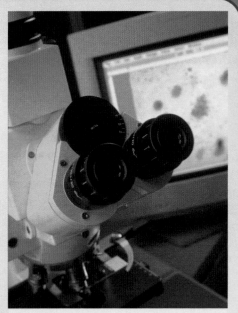

Figure 65 A digital microscope

Open your file called 'Artwork and imaging software'. Create a new bold heading '**AIS Activity 3.1**' and write a short paragraph describing what you have learned when using a digital microscope. Save the file.

Figure 66 A digital camera

Using a digital camera and downloading images

Digital cameras look very similar to traditional cameras (Figure 66), but most of them allow you to view the image on a small liquid-crystal display (LCD) screen built into the camera. As soon as you take a picture, you can view it on the screen and decide whether to keep it or to retake the shot. With a traditional camera the picture is recorded on film, so you have no idea how good or bad the photograph is until the film has been processed.

With a digital camera, light intensities are converted into a digital form that can be stored on a memory card or stick. The images can then be downloaded into the computer, viewed on screen, saved and imported into a document or printed on special photographic-quality paper. The digital images can also be taken to photographic shops and printed in the same size formats as standard photographs taken on film.

✓ **TiP**

When grouping images, if you click on an image to highlight it and then click on another image, you lose the first highlight. In order to group two or more images, click on the first image, hold down the Shift key (the up arrow under the Caps Lock), click on each other image in turn, then select Group from the Drawing toolbar.

Modifying images

Once an image has been captured electronically, it is possible to modify the image to suit your purpose exactly.

Grouping and ungrouping

Several different images can be combined to make a new image. The logo designed for Travelbug (Figure 67) was created using WordArt for the text, combined with two clip art images: the world plus an insect sitting on top!

Once the two images were positioned correctly, they were grouped together using the *Drawing* toolbar (Figure 67). The WordArt was resized, rotated to fit around the image, and the colour of the letters changed to blend in. All three elements of the logo were then grouped together.

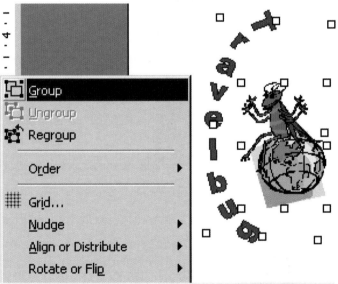

Figure 67 WordArt and clip art images are highlighted and grouped to make the logo

When the images have been grouped in this way they can be manipulated as one image. If you then wish to change part of the image, you can click on the image and this time select **Draw, Ungroup** from the *Drawing* toolbar. After the necessary changes have been made, the individual images can be regrouped to make one image. It is much easier to work with a grouped image because all the elements move together and can be resized in proportion.

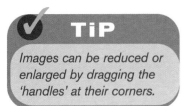

TiP

Images can be reduced or enlarged by dragging the 'handles' at their corners.

Cutting, pasting, cropping, trimming and resizing images

Sometimes you find that what you require is part of a bigger picture. For example, the Sydney Opera House scene shown in Figure 63 (page 120) included a lot of water and sky. If you wanted only the actual opera house, you could cut out that section by using the crop (trim) tool to remove the unwanted sections (Figure 68).

Figure 68 Sydney Opera House cropped and resized

Another method of cropping an image is to use the mask tools (☐ ○ ◌) that are typically available in graphics software.

- The first mask enables you to draw a rectangular shape around the area you wish to keep.
- The second mask enables you to draw a circle around the area you wish to keep.
- The third mask enables you to draw a freehand shape around the area you wish to keep.

Once the area to be kept has been identified, the image can be cropped to the mask.

To obtain the image of these mask tools, a screen print was made showing the toolbars in Corel Photo-Paint (Figure 69). The rectangular mask was used to select the mask icons and, once the image was 'cropped to mask', the rest of the screen print was removed leaving just the three mask tools.

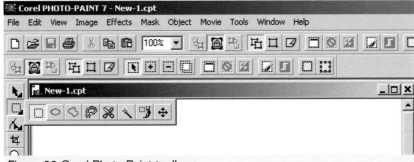

Figure 69 Corel Photo-Paint toolbars

Aligning, rotating, and flipping images

Aligning an image

The **Align or Distribute** options on the *Drawing* toolbar allow you to place an image in a particular position on the page (Figure 70).

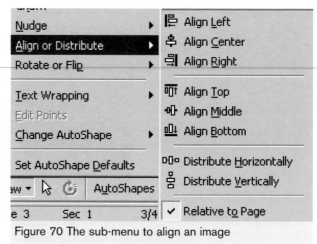

Figure 70 The sub-menu to align an image

This facility can be very useful if, for example, you wish to position the image exactly in the middle of the page, which is quite difficult just by dragging the image into position. Click on the image, select **Relative to Page** and then **Align Middle** – the image will be placed in the centre of the page.

When drawing a diagram, you may wish to draw a series of boxes the same size and then arrange them to line up evenly on the page. It can be quite difficult to do this just by dragging the boxes into position (Figure 71). Instead, select the boxes and choose **Draw, Align or Distribute, Align Left** from the *Drawing* toolbar.

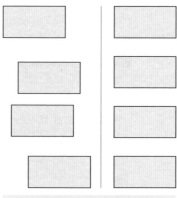

Figure 71 Boxes unaligned, and then aligned and spaced evenly

Figure 72 The AutoShape arrow rotated in various directions

Rotating an image

An image can be rotated to the left, to the right or freely. The *AutoShape arrow* points to the right, but suppose you need an arrow pointing upwards. Select the arrow and rotate left, which changes the direction to point up. If you select free rotate, then the arrow can be angled to any direction you choose (Figure 72).

Flipping an image

If you flip an image, you reverse the direction in which the image is pointing. For example, suppose you need a picture of a horse facing to the right. You have found a good picture but the horse is pointing to the left. You can use the picture and flip it to the right (Figure 73) using **Draw, Rotate or Flip, Flip Horizontal**.

Figure 73 An image flipped horizontally

You can also flip an image vertically so it turns upside down (Figure 74).

Figure 74 An image flipped vertically

Choosing appropriate image resolutions and file formats

Optimising image resolution for print and digital publications

There are two basic types of images on a computer:

- bitmaps
- vectors.

Bitmaps

A *bitmap* image is made up of dots, whereas a *vector* image is made up of various elements such as lines, curves, circles and squares.

The sharpness or clarity of a bitmap image is determined by its *resolution*, which is measured by the number of *pixels* (or dots) it contains per inch (dpi). A general 'rule of thumb' is to use 72 dpi for on-screen images and up to 300 or 600 dpi for printed images.

Figure 75(a) shows a bitmap image of a bird with a clear resolution. Figure 75(b) shows the same image increased to approximately 3.5 times the original size, and you can see clearly that the resolution has deteriorated. Figure 75(c) shows an enlarged section of the bird's beak. Notice that the edge of the beak is now looking very ragged and uneven. Figure 75(d) shows the same enlargement including the gridlines. The individual pixels are clearly visible, which is very useful if you wish to edit a picture.

 **TiP**

As a bitmap image is enlarged, the quality becomes poorer. There will be a point beyond which the quality is unacceptable for all normal uses.

TiP

If you want to create an image as a bitmap then it is important to design it at the size it should be printed.

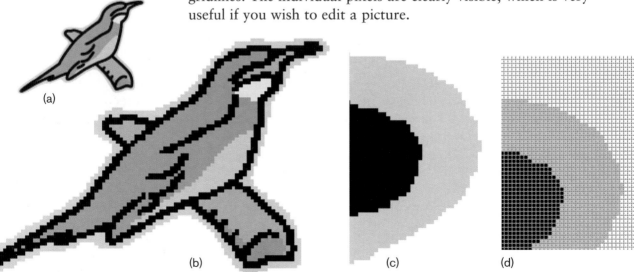

(a)

(b)

(c)

(d)

Figure 75 (a) Bitmap image of a bird. (b) The same image enlarged about 3.5 times. (c) An enlarged section of the beak. (d) An enlarged section of the beak showing the individual pixels and gridlines

Vector graphics

In vector graphics, objects are treated as collections of lines rather than patterns of individual dots. This makes it easier to enlarge the image without reducing its sharpness or quality (Figure 76). However, many people believe that bitmap images provide more subtlety in shading and texture.

Figure 76 A vector image: notice that the enlargement retains the quality

Optimising file size

Graphical images, especially photographs downloaded from a digital camera, are often very large. For example, a camera with 3 million pixels (3 megapixels) produces images of file size up to 1.5 MB, and a 5-megapixel camera may produce images more than 2 MB in size. At those sizes it can be very slow, or sometimes impossible, to download a photograph via the Internet. Therefore, it is important to think about the best way to save the file so that the file size is as small as possible without affecting the quality of the image too much.

The Microsoft Windows XP operating system includes Microsoft Photo Editor software, which can be used to resize an image. Double-click on your photograph, which will open in Microsoft Photo Editor. Select **Image**, **Resize** and change the 'units' from centimetres to pixels in the dialogue box (Figure 77). Reduce the size to 50 per cent of the original.

This *physically reduces the size of the picture*, so it appears smaller on screen. However, the quality looks just about as good, even though the file size is dramatically reduced. This will will be much quicker to send.

Another way of reducing the size of the image file is to *compress* it. Select **File**, **Save as** and then click on the **More>>** button in the dialogue box (Figure 78). Reduce the scale for the quality factor at the botton of the box. A JPEG file becomes much smaller, but the trade-off is a noticeable reduction in quality.

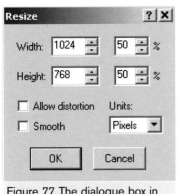

Figure 77 The dialogue box in Microsoft Photo Editor by which an image can be resized

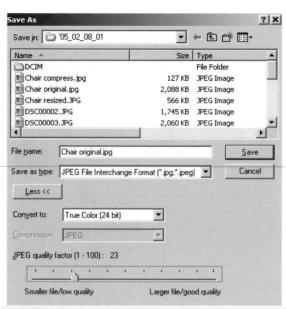

Figure 78 Compressing the file size by reducing the scale for the quality factor

Look at the photographs of a kitchen table and chair taken on a 5-megapixel digital camera (Figure 79). Note the differences in the quality of the pictures depending on the method of reducing the file size.

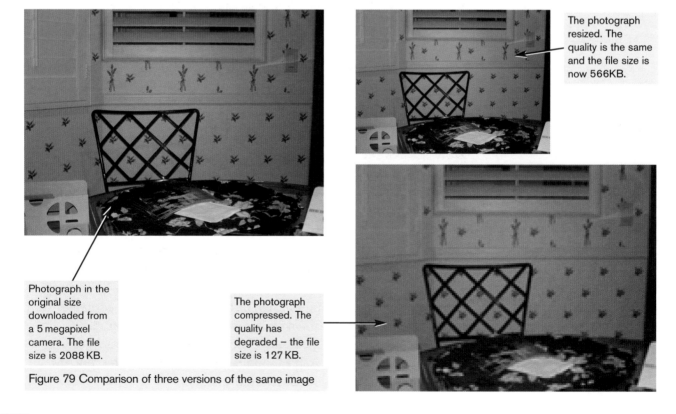

The photograph resized. The quality is the same and the file size is now 566KB.

Photograph in the original size downloaded from a 5 megapixel camera. The file size is 2088 KB.

The photograph compressed. The quality has degraded – the file size is 127 KB.

Figure 79 Comparison of three versions of the same image

Go out and try!

1 Open the file containing the photograph you took of your friend. Resize it and save it again with 'resized' included in the filename.
2 Open the original photograph again. This time compress the file size and save it again with 'compressed' included in the filename.

Open your file called 'Artwork and imaging software'. Create a new bold heading '**AIS Activity 3.2**' and write a short paragraph comparing the differences in file size and quality of the saved images. Save the file.

Optimising file formats for print and digital publications

Jargon buster

A **greyscale** image can be printed in 256 shades ranging from black to white.

The screen shots for this book were saved in TIFF format, as it supports 16.7 million colours. TIFF graphics are very flexible. They can be be any resolution, and can be greyscale or full colour. It is the preferred format for desktop publishing as it produces excellent printing results. Figure 80 shows other file formats that might be chosen in certain circumstances.

Figure 80 File formats available in this graphics software

Windows has traditionally used bitmap images (with the file extension '.bmp') for graphics. The colour quality is as good as TIFF files, but BMP files are very large, and – unlike TIFF files – they cannot be compressed. The picture of Sydney Opera House saved as a bitmap file is 1696 KB, whereas the same image saved in TIFF format was only 138 KB (Figure 81).

As the name suggests, the format devised by the Joint Photographic Experts Group (*JPEG*, pronounced J-peg) is the most suitable format for scanned photographs – full-colour photographs or greyscale images with large variations in the colour. The format is not so effective for text, cartoons or black and white line drawings.

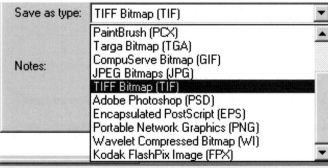

Figure 81 File size comparison

The JPEG format uses the full 16.7 million colours. Images taken with a digital camera are saved as JPEG files (with the '.jpg' extension in Microsoft applications). Image files saved in the JPEG format are considerably compressed. The Sydney Opera House saved as a JPEG file was only 20 KB.

The Graphics Interchange format (*GIF*, pronounced Jif) supports only 256 colours, which is a huge difference from the JPEG format. However, GIF is significantly better for images with just a few distinct colours, where the image has sharp contrasts – black next to white, or cartoons and animations. The format compresses images even more than the JPEG format.

Go out and try!

Open the file 'Graphic images 1' that you created for Unit 1 and change both images to greyscale. Resave the file as 'Graphic images 9'.

Open your file called 'Artwork and imaging software'. Create a new bold heading '**AIS Activity 3.3**' and write a short paragraph describing the effect greyscale has had on both images, and comparing the sizes of the two files 'Graphic images 1' and 'Graphic images 9'. Save the file.

Go out and try!

Choose any one of the images you have created in this chapter. Save it in all the following file formats so that you can compare the overall file sizes: bitmap, TIFF, JPEG and GIF.

Open your file called 'Artwork and imaging software'. Create a new bold heading '**AIS Activity 3.4**' and record the results of saving an image file in four different formats. Save the file.

Internet and intranets

In your studies you will need to research information from a wide variety of sources. A huge amount of information can be obtained if you carry out a search on the Internet or an intranet. However, be warned: there are millions of websites on the World Wide Web and, unless you have the skills to carry out an effective search, you may find it difficult to find the exact information you need.

Jargon buster

The **Internet** is the world's largest computer network, connecting millions of organisations and people across the globe.

An **intranet** uses the same technology as the Internet but is an internal communication system for a particular organisation or company. It can be accessed only by authorised users. It allows secure email communication and distribution of data.

The **World Wide Web** is a part of the Internet. Multimedia documents are connected together using **hyperlinks**. Each document is called a **web page** and a set of web pages make up a **website**.

LEARNING OUTCOMES

You need to learn about

✓ using features of browser software

✓ using search engines and portals.

Jargon buster

Broadband is the general term given to the latest in high-speed Internet access technology, which is much faster than using a dial-up modem. An Asymmetric Digital Subscriber Line (ADSL) is an example of always-on broadband technology. It uses an ordinary telephone line to allow you to access the Internet and talk on the telephone at the same time.

Using features of your browser software

The special software which enables you to search the Internet or an intranet is known as a *web browser*. A web browser enables you to view web pages and to click on links – known as *hyperlinks* – to other web pages and websites. The most common web browsers are Microsoft Internet Explorer and Netscape Navigator.

When you double-click on your browser to start it up, you can access web pages only if you are connected to the Internet via your *Internet Service Provider* (ISP). You may have an 'always on' *broadband connection*, or you may have a 'dial-up' connection (which means that your modem has to dial to your ISP before you can access the Internet).

The home page

When you launch a web browser while connected to the Internet, the default web page – the *home page* – will be loaded and appear on screen. This can be the intranet of the company whose computer you are using, the website of your ISP, or any website you have chosen.

Finding a website

If you know the *website address* (URL) you can go directly to it by typing it in the Address box of the browser and clicking on **Go**. If you don't know the address of a company or organisation, try guessing. It is often possible to guess correctly, as organisations usually try to include their name in the address. You don't even have to type 'http://www' because the browser will add that for you. Figure 82 shows the toolbar and Address box in Internet Explorer.

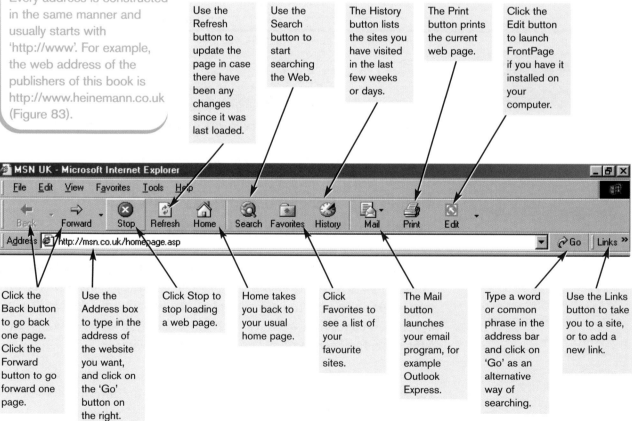

Use the Refresh button to update the page in case there have been any changes since it was last loaded.

Use the Search button to start searching the Web.

The History button lists the sites you have visited in the last few weeks or days.

The Print button prints the current web page.

Click the Edit button to launch FrontPage if you have it installed on your computer.

Click the Back button to go back one page. Click the Forward button to go forward one page.

Use the Address box to type in the address of the website you want, and click on the 'Go' button on the right.

Click Stop to stop loading a web page.

Home takes you back to your usual home page.

Click Favorites to see a list of your favourite sites.

The Mail button launches your email program, for example Outlook Express.

Type a word or common phrase in the address bar and click on 'Go' as an alternative way of searching.

Use the Links button to take you to a site, or to add a new link.

Figure 82 The Internet Explorer toolbar

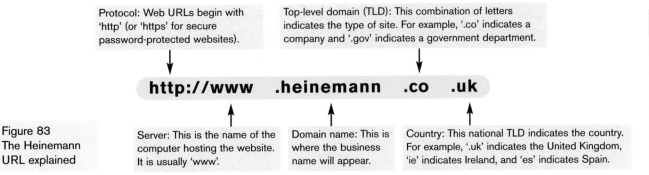

Protocol: Web URLs begin with 'http' (or 'https' for secure password-protected websites).

Top-level domain (TLD): This combination of letters indicates the type of site. For example, '.co' indicates a company and '.gov' indicates a government department.

http://www .heinemann .co .uk

Figure 83
The Heinemann
URL explained

Server: This is the name of the computer hosting the website. It is usually 'www'.

Domain name: This is where the business name will appear.

Country: This national TLD indicates the country. For example, '.uk' indicates the United Kingdom, 'ie' indicates Ireland, and 'es' indicates Spain.

Top-level domains (TLDs) indicate the type of site. Examples of useful top-level domains to know are listed in Figure 84.

.ac	A university, college or academic department
.co	A company
.com	A commercial organisation
.uk.com	An alternative area for UK registrations, often used if the .com or .co.uk name is not available
.gov	A government department
.me	An individual
.mil	A military site
.net	A network-related site
.org	Generally a charity or non-profit-making organisation
.sch	A school
.tv	The latest domain for television websites

Figure 84 Top-level domains

Hyperlinks within websites

Many companies have very large websites, starting with a home page and panning out with many pages about different aspects of their company. These pages are connected together by hyperlinks. When you move your mouse pointer over a hyperlink, the arrow changes to a pointing hand. When you click on the hyperlink you will be connected to this link.

It is very easy sometimes to get lost within a company's website, but a well constructed site will always have a link that takes you to the home page. There will usually be hyperlinks to return you back to the top of a page and to the other main areas of the site. Figure 85 shows some examples of hyperlinks.

Television

• Pick: **Himalaya with Michael Palin**
  9pm **BBC One**

• BBC TV schedules

Popular programmes:
EastEnders | Smile | TOTP | Lottery

Figure 85 Hyperlinks

Large organisations, such as the BBC, will usually include a search facility for their own site, so that you can quickly find the information you require. The BBC's website even has a 'search help and tips' page to help you find the information easily. This is a good website to bookmark in your favourites – you will find a host of useful resources to help you with your studies.

Jargon buster

Internet Explorer calls bookmarks **favorites** (spelt the American way).

TiP

If you share a machine with several people, it is a good idea to create a folder for each person. You can then keep your own favourites separate and easy to find.

Bookmarks

In the same way as you can put a bookmark in a book to return to a page quickly and easily, you can bookmark your favourite websites. You can create folders in which to file these bookmarks, in the same way as you organise and save your files on your hard drive.

Rather than having to remember the address of a website, you can easily return to a site by selecting the name from the favourites list. If there is a specific page within a website that you know you will want to return to on other occasions, you can bookmark that page, rather than the home page of the site.

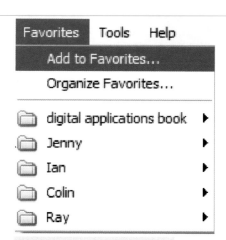

Figure 86 Internet Explorer's Favorites folder

By using the **Organize Favorites** option, you can add, rename and delete folders. You can also change the order in which you view your favourites so that websites that you visit on a regular basis appear higher in the list.

TiP

Caching pages will save a lot of time and money if you are paying for online time.

The Back and Forward buttons

All the time you are online, your browser is storing copies of the pages you have recently used in a memory area called *cache*. If you wish to return to a page you visited before, click on **Back** until you reach the page you require. Because the browser program does not have to go back to the Internet, but simply has to look into its internal cache, the page will appear more quickly than it did the first time. You can click on the **Forward** button to return to the web page you were on before you clicked **Back**.

The Refresh button

When you use the **Back** and **Forward** buttons, any changes that may have occurred since you first visited the relevant page will not appear, because the browser has not gone back to the online web page but has merely loaded that page from its internal cache. Clicking on the **Refresh** button will update the page in case there have been any changes since it was last loaded.

For example, by visiting the British Airports Authority plc's (BAA plc) website through www.heinemann.co.uk/hotlinks (express code 9842P), it is possible to check when flights have landed. When you first visit the site, the flight may be shown with its expected arrival time (Figure 87). By clicking on the **Refresh** button at regular intervals you will be able to check when it has actually landed.

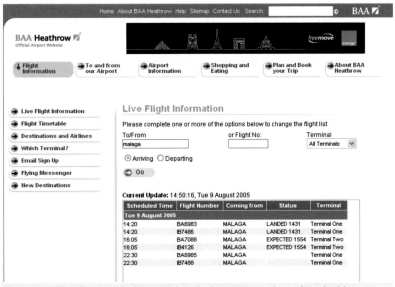

Figure 87 The BAA website, which is being constantly updated with flight information

History

Your browser will store a history of all the websites you have visited over the last few days or weeks. This can be helpful if you have forgotten to bookmark a particularly useful site. It is also useful if you have been surfing the net for some time, and you want to return to a page that you previously visited during the current session.

One way to go back to the page is to click the **Back** button, but this can be quite slow if you have visited a few pages. Another way is to use the *History* list to select the site (Figure 88).

TiP

Selecting a site from the history list is a quick way of returning to it.

Figure 88 The History list

Jargon buster

Offline refers to a way of working with web browsers or email software without being connected to the Internet. **Online** refers to a computer or other device that is connected to the Internet.

✓ TiP

Do remember when you save or copy text from web pages to take into account the copyright issues (see page 80). Some web developers protect their pages so that images and web pages cannot be copied (see Figure 89).

Figure 89 Example of a copyright message

Copying and saving web pages

You can copy text from a website and paste it into another document using the same techniques as you would in Word or Excel. You can also save whole web pages and images from a website on to your computer.

When you save a web page it will be saved in *HTML* (hypertext mark-up language). Once you have saved a web page, you can view it through your browser even when you are offline.

Selecting and printing pages

It is also possible to print web pages that you wish to keep for future reference. Many web pages, such as the one shown in Figure 90, include their own Print hyperlink. However, you can also print a page by selecting **File, Print** from the menu in the usual way.

The selected text can be printed.

Click here to print the whole page in a printer-friendly format.

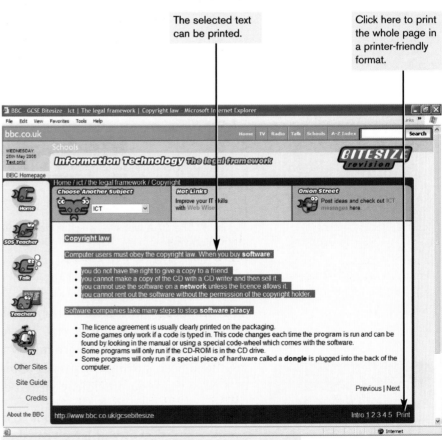

Figure 90 Selecting text to be printed from a web page

Sometimes when you print a web page you find that the material on the right-hand side is cut off. If this happens you will need to change the orientation of the page to landscape using **File, Page Setup** from the menu. You will have practised these skills already with a word processor.

If you want to print a small section of the page, again you can use the skills you have already learned when using other applications such as Word. Select the text you want to print (see Figure 90) and then choose **File, Print** and click on **Selection** before pressing **OK**.

The Go button

You can use the **Go** button on the Address bar to search for web pages. If you type a common name or word in the address bar and click on **Go**, Internet Explorer will either automatically display the web page or list those that are the most likely match.

For example, if you type 'virgin trains' in the Address bar and click on **Go** you will automatically be directed to the Virgin Trains website. However, if you type 'record shops' in the Address bar, a list of possible web pages will be displayed.

TiP

It is worth remembering that search tools are produced by businesses in competition with each other. Do not let the adverts and banners distract you from the real search results.

Jargon buster

A **portal** is a website that offers news, weather information or other services, set up as an entry point to the Web. Most portals are also search engines.

Using search engines

The Internet is the world's largest source of information. However, as you have probably already discovered, there are millions of websites, so it is sometimes quite hard to find the information you want.

If you want to find a particular company, or a certain piece of information, you will need to use a *search tool*. There are several different search tools, the main types being search engines, subject directories, meta-searches and name directories. Figure 91 explains what these are.

You might well be thinking that not only is there a huge amount of information to search, but also many different ways of searching, and you would be right! Choose one search tool and just use that one for a while. If you are not sure which one to choose, ask your friends or your teacher or tutor. When you are used to how it works and what sort of results you get then try another search tool.

- **Search engines** are indexes that work by keywords and context. Popular search engines include Altavista, Ask Jeeves and Google (see www.heinemann.co.uk/hotlinks (express code 9842P) for more). Search engines use a program called a *spider*, *robot* or *crawler* to index huge collections of Internet files. Use search engines when you want to find large numbers of related documents or specific types of document such as image files, MP3 music files or discussion lists.

- **Subject directories** are similar to search engines, but are smaller collections of Internet files grouped by subject headings. These files are not found by software, but chosen by humans. These directories are ideal if you wish to research a general topic and you want to avoid all the irrelevant files that search engines can find. An example of a search directory is Open Directory Project (see www.heinemann.co.uk/hotlinks (express code 9842P) for more).

- **Meta-search engines** send searches to several search engines at the same time. Within a few seconds, you get back results from all the search engines that were queried. Meta-search engines do not own a database of web pages; they send your search terms to the databases maintained for other search engines. You can download and install a meta-search engine to work alongside your browser. Copernic is an example of a meta-search engine; it can be downloaded via www.heinemann.co.uk/hotlinks (express code 9842P). Use meta-searches when you want to get an overall picture of what the Internet has on your topic and to check whether the Internet really is the best place for you to search.

- **Name directories** such as Yell (see www.heinemann.co.uk/hotlinks (express code 9842P) for more) are used when you want to search for people by name, telephone number, email address, postcode and so on.

Figure 91 Search tools

Successful searching

As mentioned earlier, sometimes searching can be difficult because, if you do not narrow your search in some way, you can be presented with a large number of irrelevant results. There are ways to overcome this, but these differ between search engines: look at their help pages for tips. Check out the number of hits a search produces, and, if necessary, refine your search to limit the number.

Imagine you are searching for a holiday villa to rent in the Dordogne region of France.

- If you entered *holidays* in the search box of a search engine this would be an example of a *single-criterion search*. It is important to choose the words you are searching on very carefully in order to reduce the number of results you are likely to get. This simple search produced more than 21 million results in Google, so it is unlikely that you would find the ideal holiday from using it!

- You could try typing the words *holiday villas to rent*. This *refined search* narrowed the number of hits to 668 000.
- You could reduce the number of hits further by putting double quotations marks before and after the search words. You will get more accurate information by doing so. This will find websites that include the exact phrase "*holiday villas to rent*" rather than any websites that include *any of the words in any order*. This search reduced the number of hits to 12 700.
- You can limit the number of hits further by using the + sign to show that the word must appear in the results. Typing *"holiday villas to rent" +France* reduced the number of hits to 4470. This is an example of *multiple criteria*, the first criterion being *"holidays villas to rent"* and the second being *France*.
- Some of the results included holiday villas in other countries, so typing *"holiday villas to rent" +France –Spain* will exclude all sites that have the word *Spain*. This reduced the number of hits to 1560.
- Finally, adding *+Dordogne* to the search, so that it became *"holiday villas to rent" +Dordogne +France –Spain* reduced the number of hits to 889.

TiP

Google only supports wildcards when searching for missing words in phrases. For example, "holidays in * France" *will find* holidays in Paris, France *and* holidays in rural South-West France.

Using wildcards

To make sure that you do not miss a good website by using the wrong words in your search, you can use *wildcard* matches.

You can use right-hand or left-hand wildcard searches. Entering *water** would produce results that include terms such as 'waterside' and 'waterfront'. Entering **bus* would produce results that include terms such as 'bus', 'minibus' and 'trolleybus'.

Figure 92 is a summary of search techniques.

- **Single-criterion searches** – single word searches or words grouped together by the use of quotes.
- **Multiple-criterion searches** – multiple word searches that are not grouped together by the use of quotes, or searches that are refined by using logical operators (+ or –).
- **Wildcards** – some search engines (but not Google) allow the use of * as a wildcard. For example, entering *villa** will search for sites including the words 'villa' and 'villas'.

Figure 92 Summary of search techniques

Logical operators

The earlier example of searching for holidays used two logical operators:
+ and –. Those and other operators are explained further in Figure 93.

Operator	What is does when used in a search	Example	What it means when used in a search
+ (AND)	Place this in front when you want a word to be present in the results of your search.	+Dordogne +France	The words *Dordogne* and *France* must appear somewhere in the results.
– (NOT)	Place this in front when you want to exclude a word from the results of your search.	+Dordogne +France –Spain	The words *Dordogne* and *France* must appear somewhere in the results, but all pages also with *Spain* should be excluded.
OR	Use this when you want either word to be included in the result.	France OR Spain	Web pages are included in the results if either or both search terms appear.
" "	Use this when you want the words to appear together in the same order.	"Holiday villas to rent"	The words must appear in this order.

Figure 93 The principal logical operators you will need to refine searches

TiP

Google only allows you to find synonyms (words with similar meanings) using the tilde (~) character. For example, searching for ~hotel will also return results containing the words inn, accommodation, hoteles (Spanish) and so on.

It is important to remember that different search engines will
produce different results, so it is a useful exercise to compare the
results achieved from different search engines. Most search engines
will have an advanced search option which will prompt you to refine
your search (Figure 94).

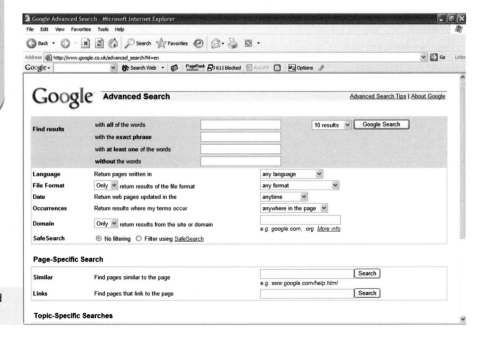

Figure 94 The Google advanced
search tool

Also, some search engines, like Ask Jeeves, will allow you to type in direct questions. For example, you could type in *Where can I find information about holiday villas to rent in the Dordogne region of France?*

Downloading images from the Internet

As well as being useful sources of information, search engines provide easy access to a wealth of images that you can use to illustrate your projects. Most modern search engines will let you search for images on a particular topic, and display thumbnails of the results so that you can easily choose the one you want.

Figure 95 shows Google's Image Search. Notice that you can switch between different types of search – Web, Images, Groups, News, and so on – by clicking the links above the text box. In Figure 95, Thomas Tripp is searching for an image of a French villa.

Figure 95 Searching for an image using Google

Figure 96 shows the first page of results. To save a particular image, click the thumbnail to go to the web page that contains the full-size image, then right-click the image and choose **Save Picture As**.

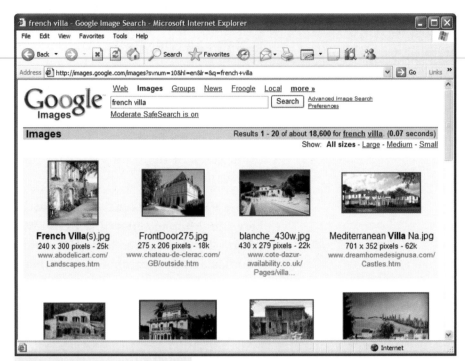

Figure 96 Image search results

Go out and try!

Search the Internet for an image of the Eiffel Tower and download it.

Open your 'Internet and intranets' file. Create a new bold heading '**II Activity 3.1**' and write a short paragraph describing the skills you have demonstrated when downloading a picture from the Internet. Add a second paragraph explaining who owns the copyright to the picture you downloaded, and anything you would need to do before you could use the picture in your own work. 💾 Save the file.

A warning!

There is no single person or organisation controlling the information that is published on the Internet. This means that the information you find may not always be correct. It may appear convincing and correct but, in fact, be completely wrong or at least misleading. Protect yourself from inaccurate information in these ways:

- Check whether there is a date on the site, and when it was last updated.
- Look to see whether it is possible to contact the site's developer. Is there an email address to contact?
- Try to work out (or at least be aware of) whether the site was developed by someone or some organisation that wants to put across its message – for example a political (sometimes an extreme political) view.

All projects have a number of common characteristics:

- clear objectives
- a definitive outcome
- a fixed period of running time
- the possibility of being broken down into a sequence of smaller tasks.

You need to learn about

✓ the questions you should ask yourself about your project brief

✓ how the success of your project will be judged

✓ how to create a project plan.

Most things you do, no matter how simple or complex, require careful planning and preparation in order to be successful.

Before you start it is essential to

- read the whole project brief carefully
- decide/understand what you have to do
- create appropriate directory/folder structures to organise your e-portfolio.

Even something as straightforward as baking a birthday cake needs planning to ensure you have the exact ingredients, the right size of baking tin and sufficient time for it to bake in the oven.

An example

Organising a holiday requires careful planning. Frankie and Sam would like to go on holiday together. Before they visit Travelbug to make their booking they must have a clear idea of where they would like to go, what they would like to do, and how much time they can afford to spend away from home.

Figure 97 Frankie and Sam's spider diagram records all their suggestions and ideas

They both have so many suggestions and ideas that they decide to write them all down to be sure everything will be taken into consideration before reaching a final decision (Figure 97).

After much discussion they agree on a two-week camping and sports activity holiday in the French Alps. Now that they have a clear goal, they can really begin the preparations for their holiday. These preparations will include

- researching the holiday market and booking the holiday
- obtaining passports
- arranging travel insurance and medical insurance
- preparing for the sports activities – perhaps by getting fit and buying suitable clothing and footwear
- buying foreign currency
- arranging for their pets to be cared for
- arranging transport to the station or airport.

By the time they leave, Frankie and Sam will have put considerable effort into the planning and preparation of their holiday, so it should be a great success.

The same is true for the projects you will be completing in order to achieve your Certificate or Diploma in Digital Applications: if the projects are not planned carefully they are unlikely to be successful!

Your project brief

A well-written brief will enable you to identify the key features of the project. Spend time reading the project brief to familiarise yourself with its contents. After studying it, you should be able to provide answers to the following questions, which will eventually form the basis of your overall plan.

What have I been asked to do?

The answer to this question will reveal the overall purpose of the project. This may be to

- carry out research on a particular topic and present your findings
- create multimedia products
- produce images and artwork
- explore and plan for a business activity.

Jargon buster

Objectives are the practical tasks that will enable you to achieve the overall aim of the project. Good objectives are said to be **SMART**:

Specific
Measurable
Agreed
Realistic
Time-constrained.

In other words, you should make sure that the tasks you set are achievable in terms of your own skills, the resources that are available to you and the time you have to complete them. Keep this in mind when you start to plan your project in more detail.

What do I have to produce?

When you are clear about the overall aim of the project, you should study the project brief more closely and make a list of all the separate items you must produce. These may be a combination of text, graphics, images or sound. For example, you may have to produce a letter, a web site, a presentation, a graphic image, a leaflet and a report. The items in your list will become the *objectives* of the project.

Why am I doing it?

Once you have established what you have been asked to do and what you are going to produce, you should think about *why* you are doing the work. This will ensure that your 'products' are entirely suitable for their purpose.

There will usually be a specific reason for producing the product:

- writing a report to convince people that something should change or remain the same

- producing a series of publications to educate people about a topical issue
- producing a multimedia product to convey a message to the public
- producing a graphic image to promote a product or event.

If you are unclear in your own mind as to the reason you are doing the work, it is highly likely that your audience will also have difficulties in understanding your message. To be successful you must make sure that your products convey the message in the most appropriate way.

Who are the target audience?

Another important point to consider is what type of people your product is directed at. In other words, who are the *target audience*? You would be expected to choose different methods for conveying the same information to children compared with adults.

For example, a poster aimed at children should be quite simple and straightforward, whereas you might use a more subtle approach for adults. If you were producing a written report for technical experts it would be quite acceptable to use technical language. However, you would have to reduce the level of technical jargon if the same report were to be presented to members of the general public.

It is important that you choose an appropriate style for your audience whilst ensuring that your finished product is also fit for its intended purpose.

TiP

Make sure you allocate some time to dealing with the unexpected. It is very unusual for everything to go smoothly, so there will almost certainly be unforeseen problems and delays along the way.

When do I have to have it finished?

Think about your timescale and how you will fit everything in. You should expect to spend a total of about 30 hours on each project. Take into account the time you will spend on planning and preparation in addition to actually producing your products.

Include time to review your work, and allow other people time to test your products. You will almost certainly need to modify or amend your work, and this must *all* be completed within the time allocated for the project.

TiP

If you wish to use a photograph that somebody else has taken, you must remember the law relating to copyright and obtain written permission to use the image. This will also take time.

Allow time to locate and order multimedia components, such as video and audio recordings from film libraries or video collections, if you need to include these in your project.

TiP

If a resource is going to be difficult to access, it might be worthwhile considering if there is an alternative way of achieving the same outcome.

Assessment Hint

You must ensure that your work complies with the technical specification outlined in the summative project brief. Make sure you follow the guidance given about file formats, download times and the overall size of the e-portfolio.

It is important that you do not take on something too ambitious that you will not be able to finish properly because you run out of time!

What resources can I use?

Each of the summative project briefs will include links to relevant websites or titles of textbooks that may help you. You will clearly have to use other resources too.

Some resources – such as school or college library books, the Internet, computers and printers – are probably readily available to you, but others may not be. For example, you may need to use specialist software or equipment that is available only at school or college, and you might need to share it with 20 or 30 other students. Alternatively, you might need to obtain books or other media from an outside source such as a library. This can take time to arrange.

If you know that something is going to be difficult to get hold of or restricted in use, you must keep this uppermost in your mind when planning your project. Furthermore, you must remember to record details of the sources of all the materials you use and to include these as evidence in your e-portfolio.

What else do I need to consider?

If you are working on artwork and images, you will also need to consider how the image is to be published. Consider, for example, whether it is to be published on screen or printed on paper. You can clearly see how an image will be viewed as part of a screen-based publication, but it is more difficult to visualise an image incorporated in a paper-based document. Carefully check the size of the image, the position on the page, and so on.

How will the success of your project be judged?

Your work will be presented in an e-portfolio that will be reviewed by your assessor and the moderator from the examining body, Edexcel. You will not be there to show them where to find the evidence of your work, so *your e-portfolio must be well structured and easy to use*. Follow the advice on creating e-portfolios given in the summative project briefs and in Skills section 5 (page 163).

You will be assessed against a number of activities, including

- planning and managing your project
- reviewing your project
- presenting evidence in an e-portfolio.

The other activities differ for each project, and more information is given in the appropriate chapters. Each activity carries a range of marks, and the marks are awarded according to the complexity of each activity.

The questions the assessor or moderator will have in mind when assessing your work are

- Is the product fit for the purpose?
- Is the product fit for the intended audience?

The safest way to make sure the assessor or moderator can answer 'Yes' to these questions is to make certain your work is reviewed throughout production and that you take notice of any points your reviewers make.

Who will review my work, and when?

It is very easy to get sidetracked and to lose sight of the original objective when you are working under pressure. A fresh pair of eyes can sometimes help to show where you have gone adrift. So for this reason you must make sure that you ask someone to look at your work at regular intervals to ensure it is fit for both the purpose and the intended audience.

You could ask your teacher, or maybe your friends or a member of your family, to help you. You could find someone who falls into the target audience category and ask him or her for an opinion.

You should show your product to a variety of different people and welcome their opinions. Constructive feedback is very valuable and is not a criticism of your work but is offered to help you improve your work and thereby achieve better marks. Listen to what other people say and consider their ideas carefully. You may not always agree, but a second opinion is always worth thinking about.

Getting started

When you have read through the summative project brief, ask your teacher or tutor to explain anything you do not understand. In this way you will avoid wasting valuable time by starting work on something that is not required!

TiP

Remember that when you start to plan your project in detail you must ensure you allow time for other people to look at your work during production as well as at the end.

You might find it helpful to produce a table, similar to the one in Figure 98, where you can make notes as you work through the project brief. It will help to ensure that nothing important is overlooked. It will also form the basis for the detailed project planning that must be done before you start work.

What have I been asked to do? This is the overall aim.		
What do I have to produce? These are the objectives.	Why am I doing it? This will help you focus on the purpose.	Who are the target audience?
1	1	1
2	2	2
3	3	3
4	4	4
5	5	5
6	6	6
7	7	7
8	8	8
9	9	9
10	10	10
When do I have to have the project finished?		
What resources can I use?		
Who will review my work?		

Figure 98 A chart to help you organise your thoughts

Project ideas

You are now almost ready to embark on your project, but you are still not ready to switch the computer on! There are a few more issues to be considered.

First you need to come up with some ideas. One of the most successful ways of doing this is to hold a brainstorming session with a group of friends – just like Frankie and Sam when they were planning their holiday! Record everything you can think of so that nothing is forgotten or overlooked.

Some people find a simple hand-written list is a good method of keeping track of ideas. One of the most popular methods is to produce a *mind map*. Mind maps help you organise your thoughts on one page so that you can see the relationship between one idea and another, and your ideas can flow from one topic to the next. Simple images can help you focus your ideas in a fun way.

Record the main topic at the centre of the page. From there, branch out and add the major themes around the main topic. Add associated ideas with further branches. Once your mind map is finished you can weigh up the pros and cons of the various ideas before coming to a decision.

This mind map (Figure 99) was used to help plan an 18th birthday party.

Be SMART (see page 146) and don't get carried away by grand ideas that go off at a tangent. Always keep the target audience and project objective in mind. Some good advice is to keep things relatively simple – think 'quality' rather than 'quantity'!

Figure 99 Mind map for planning an 18th birthday party

Your project plan

Jargon buster

A **milestone** is a checkpoint in a project, usually associated with the delivery of some version of a product. Your final deadline is a milestone, as are each of the points at which you intend to get your product reviewed.

After preparing a list of objectives from the summative project brief, your next job must be to rearrange the list into a logical sequence, remembering that some tasks may have to be finished before others can be started.

List the key milestones in each project, and, where possible, try to finish one thing before getting too involved in the next. Number the tasks in the order you will complete them.

Now look at each task in turn and break it down into a sequence of smaller tasks so that you know exactly what has to be done. Give each of the smaller tasks a number. Remember to include time to ask your reviewers to give you feedback on your work.

For example, if the first task is to prepare a web page to advertise your school/college, you might come up with a list like this:

> 1 Prepare a web page to advertise school/college
>   1.1 Plan the layout of your web page on paper
>   1.2 Write supporting text and decide on font style and colour scheme
>   1.3 Locate map of school/college
>   1.4 Take the photos and/or video clips and edit as necessary
>   1.5 Produce the web page
>   1.6 Make any changes if necessary

Decide how much time to allocate to each part of the project and write down when you plan to start and finish each task. Remember that you should also include time to deal with anything that might go wrong.

You should repeat this process for each of the objectives you have identified. The flow chart in Figure 100 can be used to remind yourself of the steps to take when planning each project.

When you have planned your project in detail you might find it helpful to record this information in table format similar to Figure 101. Ask your teacher or tutor to approve your plan before you start any work on the project. As you complete each task, tick it off. Keep a record of any changes you had to make to the plan, and say why.

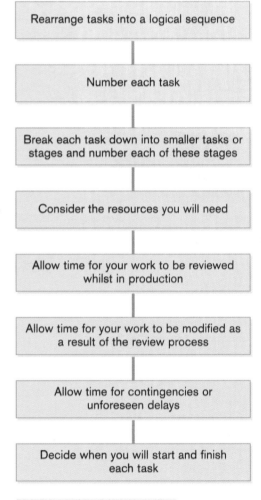

Figure 100 Planning a project

Summative Project – Unit 3		Project Start Date: 9 January			Project End Date: 31 January	
What will I produce?	How will I go about it?	How long will it take to complete?	When will I start?	When will I finish?	Done ✓	Record your progress and note any changes you make to the plan and why.
Tasks 1–3 Web page – planning and preparing	**Task 1** Plan the layout	½ hour	9 January	9 January	✓	
	Task 2 Write supporting text and decide font style and colour scheme	3 hours	9 January	9 January	✓	
	Task 3 Locate the map	1 hour	10 January	~~10 January~~ 11	✓	I thought there was a suitable map available, but when I finally located it, it was very worn out, so I had to prepare a new one. Then I ran out of time to think about the images.
Task 4 Photos and video clips	**Task 4.1** Decide which images may be suitable to include	1 hour	~~11 January~~ 12	~~11 January~~ 12	✓	
	Task 4.2 Obtain permission from the Head of Year to take photographs/video clips	½ hour	~~11 January~~ 13	~~11 January~~ 13	✓	The Head of Year was at a training event so I couldn't see her until Friday.
	Task 4.3 Take the photographs and video clips	4 hours	~~12 January~~ 13	~~13 January~~ 16	✓	This delayed me starting to take the photographs and video clips, but I knew I couldn't start the next stage until my lesson next week, so I was able to catch up as I had allowed a few spare days at this point.
	Task 4.4 Edit digital photographs and video clips in weekly lesson; make the final selection	8 days	17 January	24 January	✓	
	Task 4.5 Amend the layout of the web page if necessary	½ hour	25 January	25 January	✓	
Tasks 5–6 Producing the web page	**Task 5** Compile the web page using all the components that have been prepared	8 hours	26 January	27 January	✓	I was really pleased with this stage as it went very smoothly, probably because I had all the sections prepared before starting to create the web page, and I finished it a day early.
	Task 6 Show the web page to 2 students and 2 non-students and note comments	2 hours	27 January	29 January	✓	
Task 7 Final web page	**Task 7** Final review/edits	1 hour	30 January	~~30 January~~ 2 February	✓	I was shocked at the comments made, especially from people who didn't know the school. They found the font style and colour scheme really difficult to read, and the photograph and video clip I thought was imaginative, made no sense to them. I was so confident my web page would be fine, I didn't allow enough time for this stage, and I missed the final deadline.

Figure 101 A chart to monitor progress

Use the plan to help you meet your deadlines and finish your project on time. Every time you need to modify your plan you must save it with a new filename so that you can provide evidence of the development of your project from the beginning to the end. Don't be tempted to produce the plan after you have finished your project! Keep in mind that your teachers and the moderators who will look at the plan in your e-portfolio are very experienced and can usually discover the shortcomings!

The chart should include headings for the tasks and sub-tasks. These should be in the order in which you will carry them out, and should show how much time you will spend on each and when you will ask your reviewers for feedback. You should also include

- task descriptions
- start dates
- finish dates
- times allowed
- notes to record progress and changes made to the plan.

When you have a complete list of all the tasks, you might also find it helpful to show the proposed timing on a chart similar to Figure 102. This style of chart is called a *Gantt chart*. It helps you to see at a glance how you have spread the workload and how you can modify the plan if you run into difficulties. It is a good way of reviewing the progress of your project so you can rearrange the rest of your work if necessary. The chart in Figure 102 was produced using a table but you could also consider using a spreadsheet.

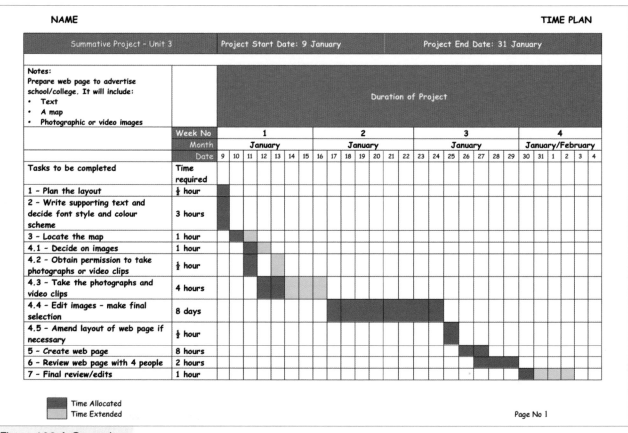

Figure 102 A Gantt chart

Assessment Hint

Don't forget to include in your e-portfolio demonstrations of how you used different software tools in your project. Screen captures can be useful for this. Press the Print Screen (PrtSc) Key on your keyboard and then paste the image into a graphics package where you can crop the image to show the relevant tools.

If you wish to take a screen print of an open dialogue box, press Alt and Print Screen together. Only the top image will be captured so you won't need to crop it.

A screen capture pasted directly into Word and cropped within Word will result in a very large file size. The file size is greatly reduced if the image is cropped in a graphics package and then inserted into Word.

Use one colour to show the time you plan to spend on each task. If necessary, you can amend it to take account of any modifications to your timescale, showing them in a different colour.

If you have access to Microsoft Works on your home computer, you may perhaps consider using the Project Planner to list each task of the project (Figure 103). You can also record the due date and make notes to remind you of what you have to do, or to record any reasons for modifying your project plan.

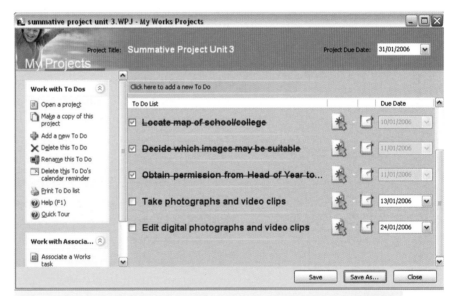

Figure 103 Microsoft Works includes this simple project planning software

Skills check

The designers of large, commercial projects often use special project management software to help them plan and monitor the progress of their work. If you study for Unit 4 you will learn to use specialist project planning software to help you develop your plan and produce a Gantt chart. You will be able to check that the progress of the work follows the plan and, if necessary, update the plan to take account of any unforeseen delays or problems. In this way the plan will become a reliable, working document that you can refer to at any time to see what still has to be done. If you work alongside the plan and keep it up to date, it will help ensure your project is completed within the given timescale.

Hints and tips

Before embarking on your project work, you will have studied the skills chapters on word-processing, spreadsheets, databases and so on. It is important that you make full use of these skills in order to create documents and on-screen publications that convey the message in the most effective way possible.

Here are some other hints to help you along the way:

- Remember to plan your e-portfolio before you start. Decide on the filenames you will use and make sure you know the appropriate format for publication in your e-portfolio. The summative project brief will specify the acceptable formats for each project. Don't waste time as you go along by converting every file into the final published format. It is a good idea to convert only the files that you will be presenting in your e-portfolio. You are less likely to get in a muddle if you do not have too many different versions of each file!

- Don't waste time in class working on things that can be done away from the classroom. Use the time you have with your teacher or tutor to get any help you need.

- Use folders and sub-folders to organise all your project work. Don't forget that you will be required to include evidence of your product from design through to implementation.

- Save updates of your work regularly with different filenames. In this way, in the unlikely event you find yourself with a corrupt file, you can revert to the previous version and you will not have too much work to repeat. This will also help you demonstrate progress in your e-portfolio.

- A USB memory stick is a good way of transferring your work between school or college and home. You will soon find the data files are too large to fit on to a floppy disk. Remember to scan your files for viruses!

- Make regular backup copies of all your data files on CD as a safety precaution. Keep the backup copies in a safe place. It is sensible to save your project work on the school or college computer system as well as at home.

When you are happy that you can put a tick against each of the following points, then you have probably done as much as you can to ensure success:

- Have you followed your plan and used it to monitor your progress?

- Have you recorded and justified any amendments you made to the plan?
- Can the history of the project be clearly seen from your plan?
- Have you kept a record of *all* the resources you used?
- Have you produced everything that you were asked to produce?
- Does your document or presentation have an appropriate layout and structure?
- Is the presentation style (font style, heading style, background colour, slide layout, and so on) consistent throughout?
- Have you used page breaks sensibly?
- Have you made effective use of space?
- Have you chosen an appropriate font size?
- Is it written in an appropriate language style to suit the target audience?
- Have you spellchecked *and* proofread your work?
- Have you used WordArt, clip art and colour in moderation?
- Are the content and images relevant and suitable for the intended audience?
- Are your images clear and a sensible size?
- Are download times acceptable for images in your web pages?
- Does your document or presentation look professional?
- Have you reviewed your work and asked somebody else to test your product?
- Have you listened to their comments and modified your product if necessary? Remember to include evidence of your modified work in your e-portfolio.
- Is the product fit for the intended purpose?
- Do all navigation routes work?
- Have you saved your work in the specified file formats?
- Does the overall size of your e-portfolio comply with the technical specification?

Keep these guidelines in mind and you should be able to produce a first-class product that will meet all the original objectives.

Good luck with all your projects!

It is important that you undertake a thorough review of your project once it is completed. You should consider

- outcomes (the publications you have produced)
- process (how you worked)
- performance (the skills you have demonstrated).

LEARNING OUTCOMES

You need to learn about

✓ collecting and presenting review evidence

✓ how to analyse the success of your project.

Introduction

Think back to Frankie and Sam and their two-week camping and sports activity holiday in the French Alps (page 144). Did you imagine that when they returned from their holiday they would forget all about it? Of course not! When they came home they would have discussed every aspect of the holiday. It is quite natural to look back on the experience, reflect on the good parts and try to work out why other aspects did not go quite so well. If they decide to repeat the holiday again some time in the future they won't want to repeat the mistakes too – they will want to have an even better time.

This is exactly the same process that you will carry out when you have completed each project. The process of reviewing your work is equally as important as actually carrying out the work in the first place. A thorough and effective evaluation will consider the *outcomes* (what you produced), the *processes* (how you produced it) and *performance* (your own contribution). You will make your own judgements and will also seek the opinions and views of a variety of other people in order to provide a comprehensive evaluation with suggestions for improvement.

TiP

It is sensible to ask several people who have not seen your work in production to give you their feedback. These are likely to be people other than your teacher and friends at school or college – for example, members of your family or people with a particular interest in the topic you have been working on.

You will have asked several people to review your project whilst you were working on it and you should have acted on their advice. You must also ensure that your finished product is reviewed by yourself and others.

Review evidence and presentation

The evidence you collect to show that you and other people have reviewed your project should be included in your e-portfolio. It may be presented in a variety of different ways.

For example, you may have provided a questionnaire for people to complete, so you could present the results graphically, drawing your conclusions from their responses. You may have received written evaluations, which could be scanned and presented as on-screen documents. Perhaps you have spoken to several people and can include a recording of your interviews. You may decide to prepare your own written evaluation, summarising the feedback you have received from other people together with your own thoughts. You might also consider producing your evaluation as a multimedia presentation or a video or audio diary. Your evidence is likely to be a combination of two or more of these suggestions.

Most things we do show that we all have strengths and weaknesses. Sometimes things turn out well and sometimes they are not so good. For example, one day we bake a cake that turns out to be rather flat, slightly burnt and generally disappointing. We show the cake to someone who has more experience, and he or she suggests that we might have used the wrong type of flour or had the oven temperature too high. We try again and next time the cake is much better.

Perhaps it will be obvious to you that some things did not go as well as you had planned and you recognise that there are weaknesses. In this case you will usually be very receptive to suggestions for improving it. However, it is more difficult to accept criticism when you think everything is all right, but do listen to what your reviewers have to say and do not be afraid to acknowledge weaknesses and to act on their suggestions for improvement.

Three important words to keep in mind when you are evaluating your work are *explain*, *justify* and *improve*. Imagine you are looking at a poster you have produced to make people aware of a

 **Assessment Hint**

The Edexcel moderators recognise that, with a limited amount of time in which to complete a project, things do not always work out exactly as planned. This in itself is not a major problem, provided you recognise where and how things went wrong, and can explain what you would do differently another time.

forthcoming meeting. On reflection, you feel the font size you used was too small and not easy to read. In your evaluation don't just say 'The font size was too small' – instead *explain*, *justify* and *improve*! You might say 'The font size was too small for the poster to be read from a distance, and as a result it was not an effective publication. I should have used a font size of at least 72 point so that people passing by were aware of the date, time and place of the meeting. In addition I could have made better use of white space to make the important points stand out.'

When you were first given the summative project brief, you studied it and found answers to a series of questions:

- **What** have I been asked to do?
- **What** do I have to produce?
- **Why** am I doing it?
- **Who** are the target audience?
- **When** do I have to have it finished?
- **What** resources can I use?
- **Who** will review my work, and when?

In your evaluation you should consider whether you achieved everything you set out to do. Do not just answer 'Yes' or 'No' in each case – *explain* and *justify* your answers. What went wrong and why did it go wrong? On the other hand, what did you consider to be particularly successful and why was it a success? Were you able to make use of knowledge and skills that you already had? Alternatively, what new skills did you have to learn?

Consider the resources you used:

- Were some especially helpful and others not so useful?
- Did you have any difficulty in finding useful material?
- Were you able to use specific hardware and software successfully?
- Did you choose the most appropriate hardware and software?

Remember to *explain* and *justify* any statements you make.

- EXPLAIN who 'tested' your product during its production stages, why you asked that particular person and what their comments were. As a result of this feedback, what did you do to IMPROVE your product. Remember your 'product' might be a document, an image, a multimedia presentation or an on-screen publication. JUSTIFY your reasons for modifying your product or not modifying it.

○ EXPLAIN why you asked for feedback from the people who looked at your finished project. Provide evidence that they have looked at your project, that you have considered their feedback and say whether or not you agreed with their comments. JUSTIFY your reasons for reaching your conclusions.

○ How would you make sure things didn't go wrong again? What could you do better next time? Think about what you could do to IMPROVE things!

Finally, consider what you have learnt from the whole experience? Was your time plan realistic? Which areas of the project took longer than anticipated and why? Were there aspects of your own skill-base that were weak? Why? How would you rectify the weakness?

A successful evaluation will consider the project from every aspect. In particular, your assessors will be looking to see that you have reviewed the outcomes, the processes and your own performance.

Outcomes

✓ To what extent have the project's objectives been met?

✓ How effective are the final products?

✓ Are they fit for purpose?

✓ How could they be improved?

✓ Does the mix of components enhance the message you are trying to convey?

✓ Is the information organised in an appropriate manner?

✓ How well do your multimedia products function?

✓ Do you think their structure and mix of components work?

✓ How easy are they to use?

✓ Are the file formats and image resolutions suitable?

✓ Does your artwork convey the intended message?

Process

✓ How well did you plan your work?

✓ Did you manage your time well?

✓ Did you meet the deadline?

✓ What, if anything, went wrong?

✓ Did you choose the right or best people to review your project?

✓ Would you arrange the project differently next time?

Performance

✓ What have you learned from working on this project?

✓ Did you have appropriate ICT skills?

✓ What additional training do you feel you need?

✓ How well did you communicate your ideas to others?

✓ How could you further improve your work?

✓ What have you learned about yourself whilst working on this project?

✓ Were you able to draw on knowledge or skills you have acquired in other subjects?

✓ Are you proud of your achievements?

Section 5

Creating an e-portfolio

You will create an e-portfolio to present evidence of your achievements. The assessor and the moderator will use your e-portfolio to judge your work, so you must make sure that it is self-explanatory and easy to use.

LEARNING OUTCOMES

You need to learn about

✓ what to include in your e-portfolio

✓ how to structure your e-portfolio

✓ how to test your e-portfolio.

What is an e-portfolio?

For most students, the idea of presenting coursework in electronic form rather than on paper is something quite new and challenging.

An e-portfolio is a multimedia stage designed to present your work. In this instance, the assessor and moderator must be able to find evidence of your achievements easily.

This is an innovative and exciting way of showing what you are able to do. Your work can come alive with graphics, animations and sound. It is environmentally friendly, saving pages and pages of paper, as well as being a very convenient way of taking your work from place to place. Ultimately your e-portfolio will contain additional information about you, your education and career and, in the longer term, is likely to be looked at by a wide range of people.

Assessment Hint

All the work you produce as evidence towards your Certificate or Diploma in Digital Applications will be presented for assessment and moderation in an e-portfolio. You will not be required to submit any documentation on paper.

You may use Microsoft FrontPage to produce your e-portfolio or you may use one of the many commercial e-portfolio systems available. Your teacher or tutor will provide you with more detailed information on the system you will be using. However, ideally, you should be able to access your work from any PC wherever you choose: at school or college or at home.

You have already looked at a wide variety of websites and are familiar with the features that make them pleasing to look at, easy to use and effective. You have evaluated their impact and appropriateness for the intended audience and looked at the combination of text and graphics, download times, and so on. Your aim is to replicate the successful features so that the assessor and moderator can easily find their way around your work. When you study each of the summative project briefs, notice how buttons and links take you from one section of the project to another, and how they also direct you to pages of hints and tips and back again to the main sections. Each one has been designed to make it easy for you to find your way around. Similarly, your e-portfolio must clearly guide users through your work.

Organisation of your e-portfolio

TiP

You will find it much easier if you plan the development of an e-portfolio alongside the planning of each project. In this way it is unlikely that important elements will be overlooked.

You must keep in mind that the people who will look at your e-portfolio will not have you at their side to help them find the evidence they will be looking for. The user interface is therefore an important element of any e-portfolio: it is essential that your e-portfolio be well organised, and structured so that anyone can find their way around it efficiently and without difficulty.

The skills you developed in Presentation software and web authoring software are equally applicable to your e-portfolio. Developing a storyboard, structure and flow charts will help ensure that suitable links are in place to make the e-portfolio easy to navigate and user-friendly.

The overall appearance of the e-portfolio is equally as important as the content. You must make sure that it is pleasing for users to look at, as well as being easy to use. Include a variety of interactive components such as buttons, hotspots, and links.

Content of your e-portfolio

The content will reflect the tasks set in the summative project briefs, so each e-portfolio will therefore be different from another. The project brief

will set out in detail what should be included, but the basic structure of each e-portfolio will be similar and will include the following

- home page
- contents page or menu
- the final work you have produced, which may include letters or reports, presentations, web pages, images, etc.
- evidence and explanations of the work you carried out, such as
  - project planning and monitoring
  - the development stages of your work – images, presentations, brochures, web pages, and so on – with comments justifying any modifications made
  - references to sources of documents or graphical components
  - storyboards, structure charts and flow charts
  - review and evaluation of the project process and outcomes
  - feedback from your reviewers and suggestions for improvement
  - bibliography
  - copyright information, including acknowledgements and permissions where applicable.

Saving your work

It is quite likely that people viewing your e-portfolio may not have access to the same software that you used to produce your documents, images, presentations and so on. For this reason it is essential that you save the work in your e-portfolio in suitable formats as outlined in the summative project brief. For example, paper-based documents are most likely to be in PDF format, whereas screen presentations such as web pages will be HTML files. Images will normally be saved as JPEGs, and multimedia presentations converted to Flash SWF format.

Accessibility and testing

When your work is finished and your e-portfolio is complete you must make sure it works properly. Do not rely solely on checking it for yourself – you know the way around your work and will not be able to provide totally unbiased feedback. Let somebody else who is not familiar with the contents look at it.

As a result of this feedback, it might be necessary to modify your e-portfolio. The important thing to remember is that it must be user-friendly so that the assessor and moderator can use it without help.

In particular you should check that

- the content is complete and includes everything detailed in the summative project brief
- the e-portfolio is clearly presented, easy to use, attractive and effective
- every link goes where it should go with no dead-ends
- the e-portfolio can be displayed correctly in different browsers (e.g. Internet Explorer and Netscape)
- download speeds are acceptable
- other people can use the e-portfolio without help.

> ✓ **TiP**
>
> *Sometimes it is helpful to give the person testing your e-portfolio a list of questions and ask him or her to find the answers. That person will have to rely on the links, instructions and explanations you have set up in your e-portfolio.*

Authentication

Finally your teacher or tutor must provide authentication to confirm that the work is your own.

Go out and try!

This is an opportunity for you to produce a small e-portfolio to display evidence of the various skills you have learnt. It will also enable you to learn how to save files in suitable file formats and to check download speeds for images.

When you worked through the other skills sections you saved separate files describing the skills you had been using. You included headings for each of the activities and wrote a short description of the skills you had learnt. Can you use the headings to provide a link to the examples described?

Task

1 Design an e-portfolio that contains a home page which will direct users to three of the skills pages, and from there to two of the files that illustrate the skills described. Make sure that some of the pages include graphic images so that you can test the download speeds of the images.

For example, the flow chart in Figure 104 would result in a home page with three distinct buttons to link to the word-processing, artwork and imaging, and internet and intranet skills files. From there, links would be set up to show the worked examples of the first two activities in each.

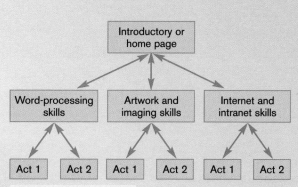

Figure 104 Flow chart

Before you create the e-portfolio, think about the file formats you should use to ensure your work can be viewed on any computer, such as using .pdf files for text pages. Ask your teacher to check that you have not forgotten anything important.

2   Ask your friends to test this e-portfolio. Consider any suggestions they make for improvement and modify your structure if necessary.

3   In turn, look at what your friends have produced and compare ideas. Think about why some e-portfolios might appear more user-friendly than others. Do you think your e-portfolio is the best possible showcase in which to display your achievements? How might you improve on your ideas when you produce the e-portfolio for your first assessment? Write a list of the improvements you might make and say why you think they are necessary – if you don't do this now you will almost certainly forget the improvements by the time you do your assessment! Save the list so that you can refer to it later.

 TiP

For practice purposes it is not necessary to include more than a basic selection of activities. The object of the exercise is to create something that is user-friendly, pleasing to the eye, easy for someone to use and viewable on any computer.

Index